Boundaries of the Universe

Boundaries of the Universe

JOHN S. GLASBY B.Sc., F.R.A.S.

Harvard University Press
Cambridge, Massachusetts
1971

Library of Congress Catalog Card Number 76-162638

SBN 674-08015-7

Printed in Great Britain

This one is for John, Anne Marie, Morag, and Raymond,
for whom this will one day be past history

Contents

Plates

Preface

For more than 2,000 years after the time of Hipparchus and Plato, astronomy was beset by myth and superstition and at best was merely a descriptive science. Inevitably, it failed to provide any real understanding of the nature of the universe. Even as recently as the mid-sixteenth century, the basic belief that the Earth was the centre of all creation, with the Sun, Moon, planets and stars revolving around it at various distances, remained unchallenged and to have suggested otherwise would have branded one as a heretic. Not until Man's true place in the universe was recognized and accepted did astronomy begin to advance out of the Dark Ages and take its rightful place among the other sciences.

Most astronomers would argue that it is only during the present century that we have gained an insight into the much wider problems of the cosmos as opposed to our more parochial view of the stellar system – the Galaxy – in which our Sun is but one inconspicuous speck among 100,000 million others. It is certainly true that in the past fifty years or so the boundaries of the universe have been pushed back at an ever-increasing rate, so much so that many astronomers alive today could scarcely have foreseen the tremendous changes which have taken place during their lifetimes.

The age of merely looking at the heavens, of mapping and cataloguing the positions of the stars down to fainter and fainter limits, is past. Throughout this long period the pace of astronomical discovery was necessarily very slow and for centuries it remained virtually static. The invention of the telescope and then the spectroscope brought about an acceleration in the rate of progress which has continued unabated to the present day. Scarcely a year now passes without some spectacular advance being made so that today, with the cooperation of modern techniques in physics, chemistry, rocket technology and biology, many of the older notions have been completely overthrown and new ideas set up in their place.

It is difficult, if not impossible, to draw any precise boundaries but it is probably true to say that chronologically the science of astronomy may be divided into three phases. First, there is descriptive astronomy, which seeks to provide us with a picture of the dimensions and masses, of the various planetary and stellar bodies, together with accurate measurements of their positions and movements. From these, the basic laws have been derived.

More recently, that branch of astronomy known as astrophysics

13

was developed, enabling us to determine the chemical and physical compositions of the stars and planets, their origins, mode of evolution, temperatures and eventual demise. Already astrophysics has made discoveries that could have resulted neither from naked-eye nor from purely telescopic investigation.

Finally, we enter the realm of cosmology which may be said to have begun with the classical researches of Edwin Hubble during the second and third decades of the present century. With the recognition that our Galaxy is but one of many millions which reach out to the very limits of the observable universe, the true position of Man in the overall concept of things became apparent.

We may feel humbled by our utter insignificance but there is no denying that we may also justifiably derive some comfort from the knowledge that, first by the use of our unfettered imagination and later by our growing scientific technology, we can range over distances which are utterly incomprehensible to the layman and back into time to the point where all of creation itself may conceivably have originated.

Coming nearer home, the rapidly developing technique of rocket research has made it possible not only for Man to go beyond the confines of this planet and set foot upon another world, but also to carry astronomical instruments beyond the atmosphere, which effectively confines our visual and photographic observations to an extremely narrow band of the electromagnetic spectrum. Now we are able to view the universe in ultraviolet light, infra-red light, X-rays, and radio emissions which are almost completely absorbed by the atmosphere that surrounds the Earth. Each problem that is solved, however, only serves to present us with a host of others of increasing complexity. The boundaries are being pushed back steadily, it is true, but the realm of the partially understood and the totally unknown is still as great as ever and it is with this vast no-man's-land of astronomy that this book is concerned.

The reader will soon appreciate that after the quiescence of the Middle Ages astronomy is now the most changing of all the sciences. This is the reason why astronomers always seem to be changing their minds over what may appear to be fundamental concepts. So many new discoveries are being made that even recent theories have to be modified, often to an appreciable extent. The arguments for and against the Big-Bang theory and the steady-state concept of the universe are an excellent example of this, while the idea of an oscillating universe has gained ground in recent years. Observational evidence in favour of all three may be put forward and it seems unlikely that a clear-cut decision will be made in the near future.

14

However, even in the midst of so much apparent uncertainty, the reader will undoubtedly experience some of the tremendous fascination of present-day astronomy – and without the unknown to provide the basic challenge of exploration, Mankind must surely stagnate.

Stevenston, Scotland　　　　　　　　　　　　　　　J. S. G.
January 1970

1. Astronomical Instruments and their Applications

The art of astronomy had its beginnings in the ages of prehistory, long before there were any written records, when men measured the lengths of the day, seasons, and the year by the movements of the Sun, Moon and stars without any knowledge of the true nature of these bodies. It mattered little to the early observers whether these various bodies revolved around a stationary Earth or whether the Earth rotated upon its axis, although the latter idea would have appeared almost incomprehensible to the early astronomers. It is perhaps inevitable that myth and superstition should have entered into astronomy and lingered for countless centuries; and it was not until the invention of the telescope that astronomy relinquished the title of an art and became a true science, subject to physical laws. Once this stage was reached and the stifling shackles of superstition were thrown off, progress was rapid, culminating in Man leaving his own planet and venturing forth to the Moon.

The main theme of the present book lies in the description of the major discoveries which have been made in recent years, covering most aspects of astronomical research; in particular, an attempt will be made to combine these into an overall and comprehensive picture of the universe as seen through the eyes of present-day astronomers. The universe is, by definition, the sum total of all things: atoms, planets, stars, galaxies, and the seemingly empty space that exists between them. It also includes Man himself and all living things, and this inevitably raises the important question whether life has come into being elsewhere than on our own planet.

First, however, we must consider the instruments used by astronomers and the techniques that have provided us with this vital information and, since certain terms will be used throughout the book with which the layman will be unfamiliar, these too must be explained in detail.

Astronomical Telescopes

Without the telescope, astronomy could never have emerged from the purely descriptive art of the ancients. The true nature of the Milky Way could never have been proved, and the existence of other vast star systems beyond our own would have been hidden from us. The ancient belief in a geocentric universe, with the Earth holding a privileged place at the centre of all things, would still have prevailed.

Astronomical telescopes are of two basic types, refractors and reflectors, the former employing two sets of lenses known as the objective and the eyepiece, and the latter a series of reflecting mirrors. The first instruments were all refractors and suffered from several defects, chiefly that of chromatic aberration. An ordinary objective does not bring the rays of light of different colours to the same focus, with the result that the image of a star appears surrounded by a coloured halo. Since chromatic aberration may be reduced, although not entirely eliminated, by increasing the focal length of the telescope, the seventeenth century saw the construction of many large and unwieldy instruments. Not until the discovery of the achromatic objective by Dolland in the middle of the eighteenth century was it possible to make refracting telescopes of more reasonable proportions. The achromatic lens consists simply of a converging lens of crown glass in combination with a diverging lens of flint glass, the opposing chromatic aberrations of the two lenses being adjusted to make the chromatic aberration of the combination negligible.

The reflecting telescope, on the other hand, using a spherical or parabolic mirror as the objective, does not suffer from chromatic aberration since in this case all of the rays are brought to the same focus irrespective of colour. One further advantage the reflector has over the refractor is that the largest lens that can be produced is far smaller than the largest mirror. There are two main reasons for this: firstly, the internal stresses set up within a large lens are much greater than in a mirror of comparable size; second, a lens must be supported by its edges which are its weakest part, this alone seriously limiting the diameter of lens that can be satisfactorily used. One further important point is that the thicker the lens, the more light it absorbs, whereas the reflective coating of silver or aluminium on a mirror absorbs very little. For these reasons, the largest telescopes are all reflectors.

The largest telescope in operation at present is the 200-inch Hale reflector at Mount Palomar Observatory, although a 238-inch reflector will shortly be commissioned at a site in Georgia, U.S.S.R. It is extremely unlikely that any larger instruments of high optical quality will be built as the terrestrial atmosphere makes it impractical to use them satisfactorily, but large flux collectors giving less sharp

18

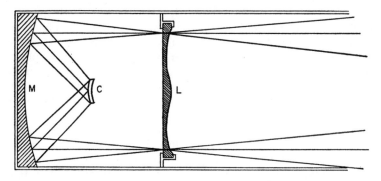

Fig. 1 Diagram of the Schmidt camera. The aspherical correcting plate L has a spherical aberration opposite and equal to that of the spherical concave mirror M. The curved plateholder C provides a sharp image over a wide field.

images are possible for specialized purposes such as optical interferometers. If any more powerful instruments of high quality are brought into operation, it appears probable that they will be erected upon the lunar surface, where atmospheric turbulence presents no problem and the lower gravity will, to a certain extent, reduce the mechanical problems of their erection.

From what has just been said, it might seem that the ordinary reflector has no serious disadvantages. This, unfortunately, is not strictly true. A parabolic mirror such as is used in the large reflectors suffers from what is termed coma, which has the effect of distorting star images in the field if they are situated some distance from the optical axis of the telescope, that is, around the edges of the field of view. This seriously reduces the size of field that is usable by the astronomer, particularly since the coma-free area decreases with increase in the size of the objective. Not until 1930, with the invention of the Schmidt camera, was this difficulty successfully overcome. The optical system of this instrument produces a coma-free field by means of a spherical mirror and an aspherical correcting plate which has a spherical aberration equal and opposite to that of the mirror (Fig. 1).

One may ask why this instrument is termed a Schmidt camera rather than a Schmidt telescope. As will be seen from Fig. 1, the image of the field is formed inside the instrument and consequently, without very drastic modifications, it can only be used as a camera and not visually. A further modification was made to this instrument in 1940 by Maksutov in Russia and Bouwers in Holland, and this is generally known as the Maksutov telescope. Since, in both types of instrument, the plateholders have curved surfaces, the plates or films have to be bent to the correct curvature before use. Nevertheless, these instruments have found an extremely important application in the photographing of large areas of the heavens down to very faint objects. Large reflectors, however, are still necessary for the purpose of flux collecting and obtaining large-scale images in an area of half a degree or less.

With very few exceptions, all of the instruments in the large observatories are mounted equatorially – that is, the telescope can be rotated about two mutually perpendicular axes, one of which points exactly to the north (or south) celestial pole. By rotation about this polar axis alone, the telescope can be made to follow a particular star with only one motion as the rotation of the Earth moves the star across the sky; for long photographic exposures, electric driving clocks are employed to keep the star centrally upon the optical axis.

Those telescopes which are employed for the measurement of stellar positions and motions are usually refractors since the quality of the stellar image is not affected by small changes in temperature as it is with a reflecting telescope. The most accurate of these instruments, known as meridian circles, are mounted between two very massive pillars supporting the horizontal axis, the telescope being able to swing in a vertical plane with very great precision. The proper motions of the stars, that is, the components of their random motions through space which are at right angles to our line of sight, are usually measured by means of photographs taken with Schmidt cameras some years apart. The photographs are compared by means of an instrument known as a blink miscroscope, a device in which the two photographs, placed side by side, are illuminated alternately. Any star that has changed its position during the intervening years will seem to jump back and forth, thereby betraying its presence. The change in position may then be measured extremely accurately with the aid of a micrometer. Micrometers are also used extensively to measure the relative positions of the components of double stars, particularly those which are physical doubles, or binaries, in order to determine the period of rotation of the components about their common centre of gravity.

Telescopes perform two main functions for the astronomer. To the layman, the chief function would seem to be that of magnifying an image, but in general this is of secondary importance as far as astronomical work is concerned. The magnifying power – of a telescope is a function of the focal lengths of the objective and the eyepiece, given by the simple formula:

20

$$M = \frac{F}{f},$$

where F is the focal length of the objective and f is that of the eyepiece. For any given instrument, therefore, we may change the magnification quite simply by merely changing the eyepiece. Of much greater importance to the astronomer are the resolving power and the light grasp of the instrument, both of which depend only upon the diameter of the objective.

Owing to the wave nature of light, whenever it passes through an aperture that limits its amount, for example the objective of a telescope, the light becomes spread out, or diffracted, resulting in an overlapping of very close images such as those of a very close double star. Since this effect is inversely proportional to the diameter of the objective, small apertures will reveal only a single image, the components being separated into discrete points of light only when the aperture is sufficiently increased. Very large apertures will therefore, not only resolve two very close stars but will also reveal finer detail upon the lunar and planetary surfaces.

The light grasp of a telescope, on the other hand, is directly proportional to the square of the radius of the objective and a large value of it enables us to see fainter stars. As a result, the aperture of a telescope is of prime importance both for resolving fine detail and for penetrating to the faintest possible limits.

Spectroscopic Analysis

By the middle of the nineteenth century, telescopes had improved to such an extent that the Milky Way had been resolved into myriads of very faint stars and the general outline of the Galaxy, the huge system of thousands of millions of stars of which our Sun is one, had been defined as a vast lens-shaped disc of incredible size. Yet in spite of this advance in our knowledge, astronomers knew nothing of the chemical and physical constitution of the stars, or of the faint nebulous objects which had been discovered in large numbers during surveys of the heavens. Indeed, it seemed impossible at the time that anything further could ever be discovered about them. The large telescopes revealed stars many times fainter than those which could be observed with the naked eye, but nothing of their composition.

As long ago as 1666, however, Newton had carried out the fundamental experiment which proved to be the basis for astronomical spectroscopy, namely, that of passing a beam of sunlight through a

21

glass prism, whereupon a band of colours – red, orange, yellow, green, blue, indigo, and violet – is formed, this band of colour being named by him the solar spectrum. Until 1802 this spectrum was considered to be continuous, with each colour blending imperceptibly into the next, but in that year Wollaston allowed the sunlight to pass through a very narrow slit rather than through a circular hole as had been done previously and found that the spectrum produced was crossed by seven dark lines. As it happened, five of these lines appeared to separate the component colours and he considered them to be merely dividing lines between one colour and the next. It remained for Fraunhofer, in 1815, to show that if a small telescope is used in conjunction with a prism mounted on a goniometer the solar spectrum contains not seven but several hundred narrow dark lines.

Over the next half century or so, several chemists and physicists studied the different spectra produced by several incandescent solids and gases, and these experiments eventually showed that three different kinds of spectra can be differentiated. The spectrum produced by an incandescent solid or liquid (in certain cases also by gases, as in the Sun) is a continuous spectrum and shows only the familiar band of rainbow colours.

An incandescent gas at lower pressures, however, gives an emission spectrum consisting of a series of bright, coloured lines on a dark background. If, on the other hand, the light from an incandescent solid or a gas at high temperature passes through a gas at a lower temperature, an absorption spectrum is formed, consisting of a series of dark lines upon a continuous background. This is the type of spectrum normally given by the stars and, significantly, the dark lines occupy the same positions as those which the coller gas would produce in emission if it were acting solely as the source of the radiation, being due to the absorption of light of these particular wavelengths from the continuous spectrum.

In 1862, Bunsen and Kirchhoff established that every chemical element produces its own unique spectrum and that every element is also able to absorb the same radiations that it emits. This not only led to the recognition of a 'reversing layer' within the solar atmosphere which gives rise to the dark lines in the spectrum but also paved the way for the identification of many of the chemical elements present in the Sun from their characteristic absorption lines.

The wavelengths of the lines produced by the various elements having been catalogued, the next phase in astronomical spectroscopy came with the application of photography. Ten years after Bunsen and Kirchhoff's pioneering work, Draper obtained the first photograph of a stellar spectrum, that of the first-magnitude star Vega, using a quartz-prism spectrograph in conjunction with a 28-inch telescope. On the observational side, three types of spectrograph have been used for the study of stellar spectra.

The objective prism consists essentially of a large prism having a small angle at the apex, which is placed in front of the telescope objective. By means of this instrument, the spectra of a very large number of stars are produced on a single photographic plate. Normally, each spectrum would be merely a point of light, but if the objective prism is placed so that its edges are horizontal when the telescope is in the meridian, the spectra are then extended north and south. The drive may then be adjusted so as to lose or gain a few seconds per hour in order to widen the spectra sufficiently for observation of the spectral lines. This method was first used by Pickering in the compilation of the Harvard Sequence of stellar spectra and is extremely useful when large numbers of spectra require to be examined in a short time but, as may be realized, both the resolution and dispersion are necessarily low. In addition, the objective-prism method is satisfactory only for the brighter stars.

For recording low-dispersion spectrograms of very faint stars a slitless spectrograph is used in conjunction with the largest telescopes, the spectrograph being mounted at the focus of the instrument. Since the sky background must be minimized, there is a limit to the faintness of the object which can be satisfactorily observed. When, however, high-dispersion spectrograms of moderately bright stars are required, a slit spectrograph is employed in which the light from the telescope is first focused on a narrow slit at the end of a collimator before passing into the prism.

Another way by which a spectrum can be formed is by means of a grating. This is simply an aluminized glass plate on which a large number of parallel lines have been engraved. The grating has two distinct advantages over the prism for stellar spectroscopy. First, the same dispersion can be achieved with a faster camera (i.e. one having a smaller focal length using a practically achievable aperture). Second, the grating provides a uniform dispersion of the different wavelengths. With a prism we find that the blue end is spread out much more than the red end, whereas with a grating all of the wavelengths are spread equally.

Classification of Stellar Spectra

As the amount of spectroscopic data increased, it became possible to classify the stars according to their spectra, a technique which has

23

had far-reaching effects in astrophysics. The first attempt at classification was made between 1866 and 1869 by Secchi, who distinguished four main classes: Type I consists of hot, white stars and Type II of cooler, yellow stars; those of Types III and IV are all red stars but show recognizable differences in their spectra to justify placing them in two distinct classes.

This system has now been generally superseded by the Harvard classification of stellar spectra, which is a continuous series with the intensities of the lines changing smoothly from the hottest to the coolest stars. At first, the sequence was labelled with letters in alphabetical order, but changes and omissions were made as more detailed information was obtained and the sequence is now W, O, B, A, F, G, K, M, N, R, S, with a special class Q which is reserved for the novae. The major characteristics of the individual classes of spectra are as follows.

Class W stars are known as Wolf–Rayet after the two French astronomers who first described them. They are extremely hot, with surface temperatures as high as 40,000°K; their spectra consist of a continuous background crossed by a large number of bright emission lines mainly of neutral and singly ionized helium in conjunction with either nitrogen (Class WN) or carbon (Class WC). Only a few Wolf–Rayet stars are known and, as we shall see later, they are very short-lived in comparison with other stars and evolve extremely rapidly.

Class O stars, like those of the preceding class, have high surface temperatures, of the order of 32,000°K, and are white in colour. At this very high temperature, the spectral lines of any metals that may be present in their atmospheres appear in the ultra-violet region of the spectrum and, since our atmosphere effectively absorbs ultra-violet light, these lines are unobservable from the surface of the Earth although they could be observed from orbiting satellites. Lines due to hydrogen are present, but the dominant lines are still those of neutral and ionized helium and of nitrogen and oxygen in which two or three electrons have been stripped from the atoms by collisions within the stellar atmosphere.

Class B stars have somewhat lower surface temperatures, around 25,000°K. Since the temperature is lower, we find that the lines of helium are those of the element in its neutral state (the lower the temperature, the more difficult it is to knock electrons out of their orbits) and even nitrogen and oxygen have only one electron removed. Stars of this spectral class are found in very large numbers

24

within the Orion nebula and elsewhere in this constellation, and for this reason they are often termed Orion stars. Very blue in colour, they are giant stars with mean densities about one-tenth that of the Sun.

Class A stars show very intense hydrogen lines in their spectra, and lines due to helium are either absent or extremely faint. Their surface temperatures are about 11,000°K and since Sirius, the brightest star, is typical of this class they are often known as the Sirian stars. They appear to predominate in low galactic latitudes, along the galactic equator. Apart from stars of Class K, they are the most numerous of all the stars. Although the surface temperature is still too high for absorption lines of neutral metals to appear, those of ionized metals are found in emission.

Class F stars are intermediate between those of the preceding class and Class G. With a surface temperature of only 7,500°K, there is a marked decrease in the strength of the hydrogen lines accompanied by a corresponding increase in the intensity of the metallic lines, those of ionized calcium being exceptionally prominent.

Class G stars, which include the Sun, show very narrow lines of hydrogen with stronger lines of ionized calcium. The surface temperature about 6,000°K, is now sufficiently low for numerous lines of neutral metals to appear in absorption. The majority are dwarf stars with a mean density about twice that of water. Such stars show little galactic concentration until one includes the very faint members of this class.

Class K stars form the largest group and are orange-yellow in colour, the brightest example being Arcturus. By now, the hydrogen lines in the spectrum have been considerably weakened, although those due to neutral metals are still prominent. The surface temperature of about 4,000°K is low enough for absorption bands of certain chemical compounds, particularly hydrocarbons, to appear. These compounds are those which are relatively stable to heat and are therefore able to survive undegraded within the atmosphere of the star.

Class M stars have surface temperatures in the range 2,000 to 3,000°K, their spectra being dominated by the fluted bands of titanium oxide which, since this molecule absorbs strongly in the blue end of the spectrum, means that they are all red stars. Both giant and dwarf stars of Class M are known, the former being nearly all variable in brightness to a certain extent. The mean density of the giant stars is only about one ten-thousandth that of the Sun; the density of the red dwarfs is somewhat greater than that of the Sun. Generally, the red giants are very distant stars. Several of the red dwarfs, on the other hand, are very close. Proxima Centauri, the nearest star of all, belongs to this class.

Class N stars are among the reddest of all the stars with surface temperatures of about 2,600°K. Like the Class M stars, they are nearly all variable and their spectra possess a similar fluted appearance, although in this case the absorption bands are due to carbon compounds and the fluting goes in the opposite direction to that in stars of the preceding class. The majority of these stars lie in, or close to, the Milky Way.

Class R stars visually resemble those of the preceding class but photographically they are quite different. The violet end of the spectrum is brighter than in Classes M or N and as a result these stars are not quite so red, although they have somewhat lower temperatures, between 1,700 and 2,300°K. The characteristic bands in the spectrum are those due to compounds of carbon.

Class S stars have extremely complicated spectra, consisting of bright hydrogen lines, many absorption and emission lines, and broad absorption bands due to zirconium oxide. Almost all of these stars are long-period variables and so far no dwarf stars belonging to this class have been discovered. Prior to 1922, these stars were included in Class N.

When the various types of spectra are arranged in a sequence such as that just outlined, it is easy to see how the spectrum of a star tells us at once a great deal about its temperature since the two are so closely related. Since the sequence represents such a gradual change from very high to very low surface temperatures, we may subdivide each spectral class still further and this is generally done by using a number from 0 to 9 as a suffix. A star of type A5, for example, lies midway in its spectroscopic characteristics between types A0 and F0. In a similar manner, astronomers use other prefixes and suffixes to denote certain peculiarities found in stellar spectra. The letter 'c' denotes that all the lines are very sharp and narrow, which is characteristic of supergiant stars. The letter 'g' has a similar connotation for giant stars, indicating that the lines due to ionized atoms are fairly strong. Conversely, the prefix 'd' indicates a dwarf star, the lines of ionized atoms being relatively weak.

To understand the reason for these differences we must examine the conditions prevailing within the atmospheres of giant and dwarf stars. As one might expect, the major difference is one of density, the atmospheres of the giants being extremely tenuous and extensive compared with those of the dwarf stars of the same spectral class and therefore of the same surface temperature. Since the intensity of radiation is virtually the same in both types of stellar atmosphere, the number of ionized atoms produced due to absorption of radiation will be similar in the two cases. In the very tenuous giant stars, however, the distances between the atoms is far greater than in the corresponding dwarf stars, and the lifetime of an ionized atom (before it can capture an electron and thereby become neutral) is much longer. As a result, the lines due to such atoms are more intense in the spectra of the giants than in those of the dwarfs.

The suffixes that are in most common use are 'e', indicating that bright emission lines are present in the spectrum; 'k', that the spectrum shows stationary lines of ionized calcium, these arising in interstellar matter lying between us and the star in question; 'n', that the lines are abnormally wide and diffuse; 'v', that the spectral characteristics are themselves variable; and 'p', that other peculiarities are present apart from those already mentioned.

Monochromatic Filters

Once the technique of stellar photography became established, it was soon recognized that there is a distinction between the visual and the photographic magnitude of a star. For some stars this difference is quite small, whereas for others it is of the order of one or two magnitudes. This difference between the visual and the photographic magnitude is known as the colour index of the star; it arises from the fact that, whereas the eye is more sensitive to the red and yellow end of the visible spectrum, ordinary photographic emulsions are more sensitive to the blue and the ultra-violet. A positive colour index is therefore indicative of a red star and a negative one of a blue star. The importance of the colour index is that, even if a star is too faint for its spectroscopic class to be determined, the colour index enables it to be placed somewhere along the Harvard Sequence.

Naturally, purely visual estimates are subjective to a certain degree and more accurate results are obtained from photovisual plates, which have been sensitized to the red and yellow wavelengths and are used with a yellow filter which reduces the intensity of the blue and ultra-violet. In order to standardize this procedure still further, Stebbins and Whitford have introduced two-colour photometry, in which the magnitude of a star is measured at about 5,000

27

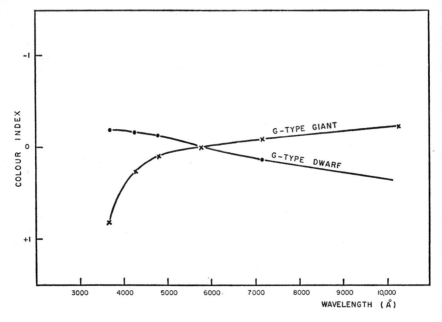

Fig. 2 Typical six-colour curves of giant and dwarf stars of the same spectral type.

Å and also at 4,300 Å, a yellow and a violet filter respectively being used. Such a method is, of course, capable of even further extension, and in the more recent technique of six-colour photometry the brightness of a star is measured in the ultra-violet, violet, blue, green, red, and infra-red regions of the spectrum. This not only provides us with an energy-distribution curve relative to wavelength, as shown in Fig. 2, but also enables us to differentiate quite readily between giants and dwarfs of the same spectral class, since giant stars are brighter in the infra-red and dwarf stars more brilliant toward the ultra-violet end of the spectrum.

When this technique is applied to the galaxies, more especially those which are sufficiently close for their internal structure to be discerned, some very interesting results have been obtained, particularly by Zwicky. The central regions of these galaxies (here we are speaking only of the spirals), are very rich in red stars and show maximum intensity in the red of the spectrum. Integration of the

28

light over the whole galaxy, however, results in a shift of intensity towards the blue and ultra-violet, owing to the much greater abundance of blue stars in the outer regions. Zwicky has employed a very ingenious technique which allows the stars of different spectral types to be identified in these spiral galaxies. Here two different prints are made; in one a blue negative is combined with a red positive, and in the other a red negative is combined with a blue positive. In the former, hot blue stars appear as black images while the cool, red stars are white; the reverse is found in the second case.

Six-colour photometry is, of course, somewhat tedious and for more routine work the U, B, V system is employed in which the brightness of the star is measured in the ultra-violet, blue and yellow regions of the spectrum. The excess blueness of the star is then given by $V-B$ and the ultra-violet flux by $U-B$.

2. The Lunar Landscape

For some thousands of years our knowledge of the universe was based solely upon what could be seen and measured by the naked eye. Unfortunately, although this yields excellent results in certain cases, it can also lead us into serious error and the early centuries of astronomy were filled with erroneous beliefs which persisted until the invention of the telescope and later of the spectroscope and other pieces of ancillary equipment now commonplace in the large observatories. The fundamental notion that the Earth was the centre of the universe died hard; many other theories have not stood the test of time and have long since been discarded.

It is important to realize, however, that it is not only the ancient beliefs which have been challenged, proved wrong, and relegated to past history. Our present concepts of the universe as a whole have altered dramatically in recent years and many ideas held only a decade or so ago have been either abandoned entirely or radically changed in the light of more recent discoveries. Astronomy, perhaps more than any other branch of science, exists in a state of continual change. Each problem that is solved, or appears to have been solved, presents in its turn a bewildering array of further puzzles awaiting solution.

To the layman, the astronomer would seem to be labouring under the most difficult and frustrating of conditions. Unlike workers in other scientific fields, he must be content merely to observe and theorize since he cannot, with rare exceptions, handle any of the objects he studies. His laboratory is the heavens, where he plays the role of a bystander unable to move far, as yet, from his native planet, witnessing a grand sequence of cosmic events beyond all imagining, striving to find the key to the secrets it contains. Where the stars and galaxies are concerned, the vast edifice of astronomical knowledge which has been built up over the centuries, is based solely upon the feeble rays of light that reach us across the tremendous vastnesses of space.

However, the universe is so vast, the time scale concerned so long, that rather than suffering from a dearth of observational data we have, more often than not, a surfeit of it. The problem then resolves itself into one of systematic analysis, of breaking down the complex information presented by the radiation we receive from the Sun, stars, and galaxies. It is with the deciphering of this bewildering mass of data that astronomers have been concerned for the past three centuries.

This book is designed to give an overall outline of our present knowledge of the universe and especially of those twilight areas which lie on the very boundaries of astronomy, out at the furthermost limits of the largest instruments. Speculation will necessarily enter into many of the arguments, for here we are often groping blindly in an attempt to understand the observations made on the very fringes of the unknown. We shall begin, however, much nearer home, with our own satellite, the Moon. For countless centuries the Moon has been an object of special attention; it is the only body on which the surface features may be examined in minute detail. It is probably true to say that, at least as far as the earthward side is concerned, we know more about the topography of the Moon than we do of our own planet.

The Composition of the Moon

Being the closest of our celestial neighbours, the Moon in all probability was formed either from the primal Earth or at some point in space very close to it. The average density of the Moon has been accurately determined; as Table 1 shows, it occupies an intermediate position within the range of densities found for the other planetary satellites of comparable size.

Table 1

Mean Densities of Satellites
(*water* = *1*)

Planet	Satellite	Density
Earth	Moon	3·33
Jupiter	Io	4·02
	Europa	3·78
	Ganymede	2·35
	Callisto	2·05
Saturn	Titan	2·41
Neptune	Triton	2·0 (?)

Here, two facts are of particular importance. Both Io and Europa have mean densities that are consistent with the presence of iron in their cores (about 20 per cent), suggesting that they are similar in composition to Mars. All of the others, with the exception of the

31

Moon, have mean densities that are too low for there to be any iron at all in their physical make-up. The Moon is, unfortunately, a borderline case. A lot depends upon the type of rock of which it is mainly composed. If this is similar to that found in the mantle of the Earth, then we may be reasonably certain there is no iron present. If, on the other hand, it is mainly of a very light kind, like that found in the Earth's crust, with a density between 2·5 and 2·8, then the Moon could perhaps contain up to 25 per cent of iron.

Here we may be tempted to ask: Why is it so important to assume that the Moon's core contains a fairly high percentage of iron? The answer lies in what has already been said, namely that the Earth and the Moon were formed very close together, at about the same time and, if they have been distinct bodies since their formation, their general composition should be similar. It seems scarcely conceivable that the Earth should have taken all of the dense elements at the time of formation and the Moon none at all.

Fortunately, there is some fairly convincing evidence that the Moon may be composed mainly of very light rock, possibly basaltic in nature. One characteristic of the Moon is that, in spite of its obviously mountainous terrain, volcanic activity is virtually absent. Geological evidence based upon the internal structure of the Earth suggests that volcanic eruptions are due to the exudation of light rock in a molten condition that is forced to the surface along pressure channels by the action of the denser material of the Earth's mantle. It may be, of course, that the temperature inside the Moon is not sufficiently high to allow of the formation of molten rock of any kind, with the result that the central regions are completely solid.

Now although we have said that there appears to be virtually no evidence for volcanic activity on the lunar surface, this statement needs some qualification. Certainly there is no positive evidence for any violent outbursts such as are commonly witnessed during volcanic eruptions on Earth, but over the years there have been many instances of changes on the lunar surface that require some comment here.

In 1903, Pickering catalogued the various descriptions of the small crater Linné that had been made by Riccioli in 1651, Schröter in 1788, Lohrmann about 1810, Mädler around 1830 and Schmidt in 1843 and 1866. This crater had been variously described as a small but extremely brilliant spot and as a deep crater, the bottom of which is totally dark. In 1866, Schmidt even found that it was not visible in his telescope although it had been readily observable with the same instrument some time before and was again visible a little while later. We could, of course, dismiss these conflicting observations on the grounds that there may have been misidentification by some of the observers, probably owing to the use of inadequate apertures, or that they did not allow for the effect of varying inclination of the sunlight, which could certainly alter the appearance of so small an object.

In 1956, however, Alter photographed Linné in violet and infra-red light, using the 60-inch reflector at Mount Wilson, with some very interesting results. Whereas in violet light Linné appears as an elevated crest on the lunar surface, in infra-red light it seems to be made up of a cluster of small craterlets. These observations would tend to confirm the earlier descriptions of this region and indicate that it may well be the centre of small, but definite, changes on the Moon.

Perhaps the most celebrated instance of activity on the lunar surface is that of the distinctive crater Alphonsus, which was also photographed by Alter in 1956. He found that in infra-red light there are small clefts around the periphery of the crater and also within the central regions, these showing up very clearly on the photographs. In violet light, however, these clefts are extremely indistinct, almost as though they were, at that time, covered by a misty obscuration indicative of some sort of vapour emission from this area. Two years later, in 1958, Kozyrev studied Alphonsus spectroscopically and announced the presence of carbon vapour rising during a thirty-minute period from some point close to the summit of the central peak within the crater. Subsequent visual studies of this region have revealed the presence of a darkish patch very close to this particular spot near the mountain crest, the colour being described variously as orange or red. It must be emphasized here that, although the spectroscopic evidence appears to be beyond doubt, certain observers have failed utterly to see the coloured patch on the side of the mountain peak. A lot, of course, depends upon the angle of the incident sunlight, and this may be a major problem in detecting minute differences in colour.

It had been hoped that more direct evidence concerning the internal structure of the Moon could be obtained from a seismometer left on the lunar surface in the Mare Tranquillitatis by the members of the Apollo 11 mission. Signals transmitted back to Earth by this instrument have recently been analysed. They have shown the presence of horizontal waves travelling through the lunar surface but, significantly, no body waves moving radially towards the surface from the interior have been detected. The possibility that these vibrations may have been due to the fall of a meteorite on to the lunar surface had been considered but virtually ruled out for several

C

reasons. The amplitude of the vibrations shows that such a meteorite must have weighed at least a ton and such bodies are extremely rare. The fact that three such waves were recorded makes it much more likely that these moonquakes were due to internal causes.

The dispersion of the signals – a progressive lengthening of the wave forms – and the absence of any vertical waves are indicative of a layer structure of the Moon. Preliminary examination of the signals indicates that they are consistent with this picture, but a more recent report has shown that no further signals have been received from the seismometer and some doubts have been expressed concerning the validity of the inferences drawn from analysis of the original signals. However, a further instrument was landed upon the Moon by the Apollo 12 mission and three more, designed to act simultaneously, are scheduled to be landed by the next three Apollo flights. If all function satisfactorily, we may confidently expect to obtain a much more detailed picture of the internal constitution of the Moon.

We cannot leave this question of the seismological records obtained from the Moon, however, without mentioning the very important experiment carried out by the members of Apollo 12. During the ascent stage from the lunar surface, the Intrepid section of the lunar landing vehicle was ejected back on to the Moon to to simulate a meteoritic impact. The results were completely unexpected. Reverberations continued for almost an hour, the effect being similar to that produced when a bell is struck. So far, only a preliminary analysis has been made of the signals but, from what has been done, it would appear as though the Moon has a peculiar layered structure.

The Lunar Craters

We now come to a very important and much-debated problem – that of the origin of the lunar craters. That these are not distributed evenly over the lunar surface is immediately obvious even in a small telescope, and observation shows that there are far more of them towards the region of the south pole than in the northern hemisphere. One further obvious feature is that they cover a very wide range of sizes, from only a few metres across, as shown in the early television pictures transmitted back to Earth in the few seconds before the unmanned rocket probes crashed and on the more recently obtained photographs taken by the manned flights around the Moon, to giant craters hundreds of kilometres in diameter.

The maria, too, are noticeably devoid of large craters, and any explanation of their formation must clearly take this heterogeneous distribution into account. We also have to explain the appearance of the craters themselves. Some, for example, have well-defined central peaks; in others these are conspicuously absent. The walls of certain craters exhibit a markedly terraced structure; others have walls that have been broken in places by smaller craters and in almost every case it is the small crater which is intact, strongly suggesting that this smaller crater was produced later than the large one. Finally, there are the large ray systems associated with certain craters which, in the telescope, give the appearance of powdered material ejected from the main crater by some form of impact, although an explosive origin of the rays cannot be ruled out.

There are two main theories concerning the origin of the lunar craters: one is that they were formed by the impact of meteoritic bodies crashing on to the surface, the other that they are due to some form of vulcanism. The first theory was advanced in 1892 by Gilbert and subsequently enlarged and modified by Baldwin. The large majority of the craters are, according to this theory, believed to have been formed during the early stages of the Moon's career when, as seems possible, the number of such meteoritic bodies moving in orbits similar to that of the Earth-Moon system was quite large. Whether the maria, the very large plains, were also formed by impact, but in this case by bodies similar in size to the asteroids, is problematical. That large numbers of such bodies would be present in the newly-formed planetary system seems very likely and, since the Moon possesses no atmosphere, such bodies would reach the lunar surface, not only with high velocities, but also would suffer no frictional burning such as they undergo when plunging through the terrestrial atmosphere.

Now what will happen when a large body strikes the lunar surface? Since it will be travelling with a speed of several kilometres per second, it will not be stopped at the moment of impact but will penetrate to a depth dependent upon its mass and impact velocity. With the larger meteorites, this may be some kilometres below the surface. The impact theory has to face several criticisms. For example, if we are to explain the very large craters on the assumption that several meteorites fell close together (in space, although not necessarily in time), this would imply that such craters should have uneven floors, not only indented to varying degrees by the several impacts but strewn with boulders and debris from the breakup of the meteorites themselves. Observation, however, suggests that the floors of large craters are smoother than can be accounted for by this theory unless there is some other mechanism whereby the floor becomes appreciably smoother following the impact.

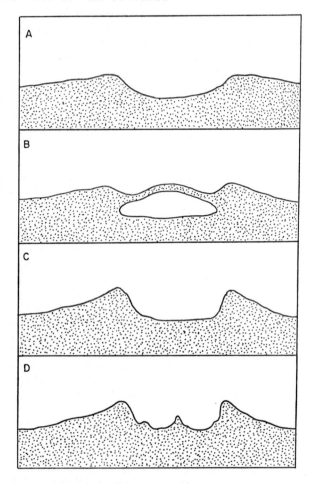

Fig. 3 Formation of lunar craters on the meteoritic-impact theory. The gas bubble eventually works its way through the surface, followed by collapse and possibly smaller secondary outbursts.

One mechanism that has been suggested by the proponents of the meteoritic-bombardment idea is that those meteorites which formed the craters fell upon the lunar surface during a period when the Moon was in a semi-plastic condition or, if they fell after the lunar crust had solidified to a large degree, that the impact itself would liquefy the surface rock sufficiently to force it to flow rather like lava, thereby accounting for the relative smoothness of the crater floors. This liquefying of the rock at the moment of impact has also been put forward to explain the presence of those craters which have a central mountain peak, since we know, from high-speed photography of drops falling into liquids that such a jet is formed in the centre of the impact region.

There is, unfortunately, one very serious objection to this idea. When a body the size and shape of a meteorite strikes solid rock it is quite probable that no intermediate liquid phase is ever produced. Rather, the rock is transformed into gas at a very high temperature and, owing to the high velocities associated with such meteorites, this bubble of gas will be forced deep into the lunar crust ahead of the downward-plunging meteorite. A little reflection, however, will show that this may not be such a serious objection as might be supposed. Sooner or later this mass of gas will force its way back to the surface, perhaps with explosive results. By this time much of the overlying material will be in a finely pulverized condition and, once the explosive force is spent, a crater similar to those now found on the Moon will result. The probable sequence of events is shown in Fig. 3.

The impact theory must also face up to several other criticisms. For example, it does not satisfactorily explain the many central peaks that are found within certain of the larger craters. Again, we find many instances where a small crater has formed along the wall of a larger one and very often the wall of the large crater is unbroken up to the wall of the smaller one. On the theory that these two craters were both formed by the impact of meteorites, we would expect the wall of the larger one to be broken for some distance on either side of the smaller crater. It appears to be stretching coincidence a little too far to postulate that in every case the smaller body fell last, as would have to be the case. We may argue, of course, that in those instances where the larger body fell last it would totally obliterate all traces of the smaller crater, and indeed there is some evidence of this in certain areas of the lunar surface.

One further piece of evidence that has been put forward in favour of the impact theory is that certain large, well-defined craters exhibit what are known as ray systems. These, unlike most other lunar features, are seen to their best advantage during the time of full moon and take the form of long, bright streaks which originate from a particular crater, often extending for hundreds of kilometres in all directions, crossing craters, mountain chains, faults and similar topographical features without any obvious break or change in direction. Two craters that are the centres of very conspicuous ray

37

systems are Copernicus and Tycho, yet a closer examination shows certain differences between these features. Copernicus is a large crater, almost perfectly circular, whose walls show a remarkably terraced structure (see Plate 1). The rays, it will be noticed, are fairly diffuse in nature and form a complex system extending over the surrounding plain and then progress across the Carpathian Mountains to the north. Tycho, on the other hand, with its well-defined central peak, does not exhibit this terracing within the walls and the rays are perfectly straight and stand out well against the background.

The photographs taken by the Lunar Orbiter rockets have been extensively studied, since they cover the site of the Surveyor VII landing. It is important to note that this area is composed of material which has clearly flowed like several liquids having somewhat different properties. Such liquid flows cannot be seen on photographs taken from Earth. The northern rim of Tycho shows, in addition to these flows, small lakes which are rather like tiny maria that appear to be volcanic rocks thrown up over a considerable length of time after the main crater was formed. Kuiper has suggested that Tycho was formed by meteoritic impact which then triggered off a series of volcanic eruptions producing these liquid flows and lakes. Green, on the other hand, believes that Tycho itself is a caldera, formed mainly by vulcanism. A third possibility is that the liquid flows are of much more recent origin than the crater and were brought about by moonquakes which caused a downward movement of material. Several similar features are known on Earth, but we must be cautious in interpreting the lunar flows in a similar manner, since water was a major factor in the production of terrestrial flows and it does not seem likely (although it still cannot be definitely ruled out) that water performed the same function upon the Moon at some time in the past.

Pictures taken by Surveyor VII show an area littered with small blocks, which have been interpreted as the remains of a fragment of debris that fell in the region from an impact crater some distance away, particularly since these rocks are lying on the surface and have not penetrated to any extent. Certain small rocks closely resemble those found near the large meteor crater in Arizona and appear to favour an impact origin for Tycho.

Copernicus is much older than Tycho, as is shown by the fact that there are far more small craters both in and around it. The outer rim, however, again shows several of the lakes already mentioned in the case of Tycho, which show up well on Lunar Orbiter V photographs (see Plate II). There also exist some liquid flows along the outer rim of Copernicus, although these are fewer than in the case of Tycho. Copernicus, however, has many flows within the crater itself, the floor of which is much smoother than that of Tycho, and the hills that are present appear far more rounded and subdued.

So far, the pictures that have been obtained either from the Lunar Orbiter rockets or from the Surveyor soft landings do not provide us with a clear-cut means of deciding for or against the two rival theories of the formation of the craters, and it seems obvious now that both means have contributed to their formation. With regard to the igneous, or volcanic, theory of their origin, there are two possible mechanisms we must consider. The first is the actual eruption of molten lava, similar to the volcanic outbursts experienced on Earth. Although in general the craters present on the lunar surface are larger than any terrestrial ones, there is often a very close resemblance to the wide, circular depressions formed by vulcanism on Earth and known as calderas. The molten lava forces its way through the crustal mass and on cooling forms a ring-like structure through which subsequent lava outflows may penetrate. There is also an associated process known as degassing, which is the release of volatile products through the same vent. Many of the observed features of the craters may be explained simply if we assume that at some time during its early life there was water on the Moon. Certainly it is unlikely that there is any free water there now (although certain transient lunar phenomena have been explained on the basis of hoar frost within the very deep craters and rifts), but if any did exist during these remote times we would expect to find evidence of those elements now present in water on Earth, namely chlorine, bromine, iodine and salt. Following the degassing process, the central portion of the crater would subside, although minor eruptions would probably occur, resulting in some of the fault-like structures found along certain crater floors (Fig. 4).

As we have already seen, the floors of certain craters, notably Tycho, show clear evidence of a liquid phase having been present at some time during the early formation of the crater, and the large number of lakes around the rims of the large craters may also be explained on the volcanic theory. The floor of the crater Aristarchus, too, presents many puzzling features which can best be explained on the basis of vulcanism. The Lunar Orbiter photographs clearly show many low, rounded hills which, although it has been suggested that they are due to some rebound mechanism shortly after the crater was formed by meteoritic bombardment, seem to be of a volcanic nature. The many crevasses that intersect the floor of Aristarchus do not appear to affect these low hills, suggesting that

39

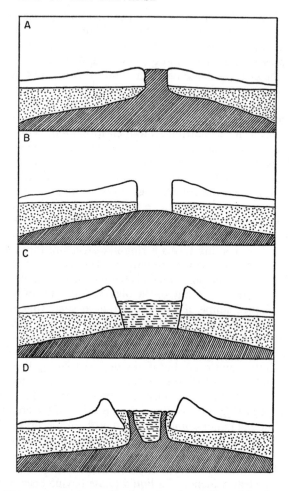

Fig. 4 Lunar-crater formation by vulcanism followed by degassing.

they are of relatively recent origin and certainly not as old as the crater itself. The most plausible explanation at the moment is that they represent extrusions of some viscous material, formed a long time after the crater and partially covered by ash. One further anomaly about ray craters such as Tycho, Copernicus and Aristarchus is that they cool more slowly than other craters, as was shown in 1964 by Saari and Shorthill, who examined these regions

40

in the infra-red during the total lunar eclipse. In addition, they reflect radar signals more efficiently, and both of these observations may be explained by the greater area of exposed solid rock, indicating that erosion has taken place to a far lesser degree here than in other craters.

The Ray Systems

The fact that the rays emanate from certain of the large craters suggests that they may be merely lines of least resistance in the surface brought about by the volcanic eruptions that formed the craters, the rays being simply wide cracks which are filled with deposits brought up from below the surface. On the whole, examination of the Lunar Orbiter photographs does not support this idea. Loewy and Puiseux have expressed the view that the rays, although produced by volcanic ash, were deposited in their present position by air currents at a time when the Moon still had an atmosphere.

So far we have assumed that the Moon has been airless since its formation. Whether or not a planet or satellite has an atmosphere depends upon what is known as the limiting speed of molecules and to a certain extent on the surface temperature. The limiting speed of molecules is governed by the gravitational field of a body and therefore by its mass. As far as the Earth is concerned, this limiting velocity is seven miles per second and any atoms or molecules in the atmosphere travelling in excess of this velocity will escape into space and be totally lost. Fortunately for the continuation of life on this planet, only the lightest of gases – hydrogen and helium – attain velocities higher than this. If the Moon did originally have an atmosphere, it would soon have been lost since here the limiting speed of molecules is only 1·5 miles per second and even the heavier gases such as oxygen, nitrogen and carbon dioxide have mean velocities higher than this.

Accepting for the moment that there may have been a short-lived atmosphere on the Moon many millions of years ago, it is difficult to understand why the ray systems should have retained their original form if they were laid down by air currents. Unless we say that the lunar atmosphere dissipated extremely rapidly, we would expect such fine dust formations to have become smoothed out in a fairly short period by such atmospheric currents.

Dome Initiation of the Craters

The second mechanism whereby we can explain crater formation on the lines of the igneous theory simply supposes that shortly after its formation the lunar surface had a plastic-like composition in

41

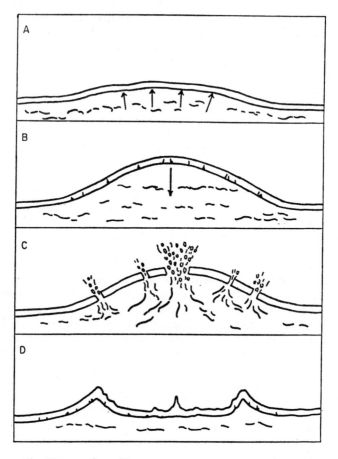

Fig. 5 Formation of lunar craters by swelling and eventual collapse of dome like structure.

which domed bulges were produced by the build-up of internal gas pressure. Such swellings would be unstable and collapse would occur in the central regions, as shown in Fig. 5. After the collapse only the outer rim would be left standing, just as we find now in a typical lunar crater. One important point about this particular hypothesis is that any actual eruption will occur in the central region of the dome, thus explaining the presence of the mountain peak that is a characteristic feature of many craters. It is, of course, unlikely that all of the gas will be released during the initial swelling and subsidiary eruptions can thus take place during later stages in the development of the crater, possibly accompanied by an outflow of magma. Whenever this happens, we will get the sort of terraced structure we find around the circumference of such craters as Copernicus.

Although both the meteoritic-bombardment and the volcanic theories are simple to apply in theory, any definite pronouncements in favour of either must remain in abeyance until the samples of lunar rock brought back by the Apollo II mission have been thoroughly studied. It must also be borne in mind that these samples were taken from only one specific point on the lunar surface and we cannot be dogmatic about crater formation simply from these few samples. The presence of basaltic material has already been confirmed and it is possible that certain of the minerals present on the Moon may contain water of crystallization. This implies that at some stage there has been free water on the Moon, although it is not necessarily still present in this form. The most that can be said at present is that both ideas account for many of the peculiarities of the craters and clearly both theories are applicable. That some of the lunar features were formed by the impact of meteoritic bodies and others by some form of vulcanism is beyond doubt. The major problem still outstanding is to determine which of the two mechanisms was the predominant one and whether either is still operative to any appreciable extent.

The Lunar Maria

Before the invention of the telescope, the large grey patches visible to the naked eye on the lunar surface were considered to be seas and hence are known as maria. Even for some time after telescopes were first used, these early instruments did not provide sufficient magnification or resolution for observers to distinguish the true nature of the lunar surface.

Broadly speaking, the maria may be divided into two classes. The first are roughly circular in outline, more often than not bordered by long mountain chains with the mare material ending at fairly well-defined limits. The second are less well defined, generally merging into the surrounding features, the mare material often being found between the bordering craters and mountains.

Through the telescope, the maria appear as vast, flat plains at a somewhat lower level than the cratered regions, but this is misleading; there are deep depressions within the maria which cannot readily be discerned since the slopes are so gentle. Only at the terminator are they obvious when the oblique rays of the Sun show them up to their best advantage. At this phase, too, it is possible to make out long ridges lying across the maria, often running parallel to the edges.

Clearly, when we come to examine the origin of the maria, we must take into account their large dimensions. If we postulate vulcanism as the cause of these regions rather than supposing that they were formed by the impact of extremely massive bodies, then the assumption is that they are large seas of lava which cooled and formed a thin crust that would seal off the rest of the molten material beneath the surface. Owing to this sealing effect, the residual lava would take far longer to cool and crystallization could then occur, initially at the lowest levels and then progressively through the mass towards the surface. Since this crystalline material will occupy a smaller volume than the original amorphous lava, there will be a gradual subsidence of the crust, again resulting in the formation of fault lines along the surface. Crystallization would also tend to trap pockets of gas, particularly around the edges of the maria, and this gas would escape either through already existing cracks in the surface or by making cracks. Some confirmation of the igneous process comes from the presence of such fissures – or rilles as they are called – around and across the maria. A further observational feature supporting this theory is that the mountain chains and also the various isolated mountain peaks which are found on the lunar surface rise abruptly from the surrounding terrain and also appear to be of a somewhat different colour, suggesting that they were indeed formed by the upthrusting of material from below the crustal zone.

When we examine the maria closely with large telescopes we find that the surface is far from smooth and featureless. Rather, it possesses a curious structure made up of large sections having a polygonal shape bounded by reasonably straight edges, the intersections being of a whitish colour. These may represent stress lines which were the cause of the early fissuring; extending this idea further, we may also postulate, as we have already seen, that the ray systems are simply the lines of least resistance in the surrounding maria.

There are a few exceptionally large maria, however, which are difficult to explain on a volcanic basis. One of these is the Mare Orientale (see Plate III), which is not only the largest of all these features but probably the youngest, although it may still be several hundred million years old. This curious feature shows several similarities to both Tycho and Copernicus, and clearly was not formed by either a simple or a single process. From its size, it has been suggested that it may have been initially formed by the impact of a comet or asteroid, and it is perhaps significant that there is no central peak. The concentric rings of mountains bear a striking resemblance to the terraced structure of Copernicus, although on a far more magnificent scale. Certainly an impact alone could not have produced a feature such as this, and the most plausible explanation is that it was initiated by impact, which was followed by both faulting and then uplift, with some flooding of mare material. Being among the youngest of the maria, Mare Orientale provides us with a good idea of what the other maria probably looked like many millions of years ago, before erosion and extensive flooding took place.

The Lunar Rilles

Mention has already been made of the lunar rilles which are found extensively intersecting the crater floors. These are large cracks in the surface which are sometimes as much as several hundred kilometres in length and are generally fairly constant in width (see Plate IV). Several theories have been advanced to explain these features. For example, they may be lava drainage channels, an idea that has received some support after the discovery of their basaltic composition by the Surveyor spacecraft. Cameron has suggested that they may be volcanic eruptions formed of small fragments and a fluidized type of ash; there is also the possibility that they may be exactly what they look – dried-up channels that have been eroded into the rock by running water.

The last explanation seems improbable, since dry water courses are known on the terrestrial surface and not only do these not resemble the lunar rilles but they also require very long periods of erosion by water. Observation shows that the large majority of the rilles begin at a crater and proceed in a linear course away from it, and it appears more likely that they are really subsidence features brought about by gaseous eruptions along fractures in the lunar surface.

Lunar Dust

Until very recently, it was widely believed that regions such as the maria might be covered by a relatively thick layer of dust, and clearly it is important for us to know the depth of dust present in these regions since they are the most suitable for further manned landings. There are several processes taking place on the Moon's surface which could result in the formation of dust, apart from that which may have been formed during volcanic eruptions. Some would almost certainly be present during the time of the Moon's formation according to the latest theories of the origin of the various bodies within the solar system. In addition, the action of the intense ultra-violet radiation from the Sun upon the rocky surface would result in dust formation, while the wide range of temperatures prevalent upon the lunar surface during the alternating periods of night and day would produce an endless cycle of expansion and contraction that would flake off small particles of rock. Since there is no atmospheric turbulence to scatter such dust, it would gradually drift, because of gravity, towards the lowest levels on the Moon, namely the maria and the floors of the craters.

One piece of evidence that appears to favour the idea of a thick mantle of dust on the maria is the presence of several 'drowned' craters found within them. These are large craters which do not stand out from their surroundings, with low ringwalls which are often not completely formed. The evidence gathered from the Ranger and Apollo orbital flights, however, demonstrates that any dust layer that may be present in these regions is only a few inches thick at most; this data eliminated the idea that the maria contain just as many craters as other regions of the lunar surface, such craters being drowned by dust so that they are no longer visible. The final proof was provided by the Apollo 11 landing in the Mare Tranquillitatis. The dust in this region is certainly no more than an inch or so deep.

How, then, can we explain the presence of the 'ghost' craters found within the maria and appearing merely as slight elevations above the surrounding terrain? If we suppose that in these regions either there was insufficient molten magma present beneath the surface to form high crater walls or the crust did not allow fissuring to a large enough extent, then it is to be expected that crater formation would take place only to a very limited degree, producing only partially formed ringwalls and very low ridges.

The Lunar Mountains

The lunar mountains are of several different types but, apart from the ringwalls of the craters and the fact that relatively little erosion has occurred on the Moon, they all find their counterparts on Earth. Sometimes they occur as isolated peaks such as Pico, but more commonly they are distributed as long chains and massifs, these being the highest mountains present on the Moon.

The isolated peaks are generally around 5,000 to 6,000 feet high, and although we might expect, from a study of the shadows thrown by them under oblique illumination from the Sun, that they are bounded by steep slopes, this is not normally the case. More often

46

than not they are appreciably rounded with fairly flat summits, as may be seen telescopically along the very limb of the Moon.

It is in the region of the south pole that we find the grandest of the mountain chains and massifs. The Leibnitz Mountains, for example, rise to a height of around 30,000 feet and are almost certainly the highest points on the whole of the visible side of the Moon. Other mountain ranges, such as the Doerfels, are more than 20,000 feet in height and there are many containing peaks that rise to between 9,000 and 15,000 feet.

The massifs are composed of a large number of mountains arranged apparently in a chaotic mass, which appear to possess no recognizable consistency in their structure. When examined more closely, however, it can be seen that there is often an underlying pattern to the massifs, which seem to be vast areas of the lunar surface that have been uplifted along one edge above the surrounding level. The Apennines, for instance, which form a border along the Mare Imbrium, exhibit a very steep edge that rises abruptly from the adjoining plain. Often, too, there are well-defined ranges running through the massifs, giving some form of regularity to their general appearance.

The Existence of Water and Air on the Moon

Earlier, we saw how the presence of water on the Moon at sometime in the remote past would lead to the formation of certain mineral deposits, particularly around the edges of the maria and possibly also the larger craters, and how the rays that cover large regions of the lunar surface may have been brought about by wind-borne dust from volcanic eruptions on a considerably larger scale than those known at present on Earth. We must now examine the question of the presence of both water and an atmosphere on the Moon at the present time.

The early notion that the maria were seas has long since been discarded. Even though we can see minute detail on the lunar surface with amazing clarity, there is nowhere any evidence for the presence of water. Admittedly, some of the latest photographs taken from the Apollo spacecraft at a distance of only nine miles above the surface show narrow valley formations that bear a striking resemblance to ancient watercourses on the terrestrial surface, but it is far more probable that they are simply faults or fissures.

Before dismissing this idea altogether, however, it is perhaps worthwhile to examine more closely the possibility that the maria are sea bottoms, the water that originally filled them having either evaporated or drained away into the lunar interior. We can perhaps

47

best do this by visualizing what the beds of the oceans on Earth would look like if the seas themselves were no longer present. From hydrographic surveys that have recently been carried out we can form a fairly accurate picture of what the ocean beds look like. They will be covered with layers of deposits carried down from the bordering land masses together with various muds of organic origin, and they will also be far from smooth, particularly near the coastlines where steep shelves are known to exist. Since these muds and oozes have different origins, we would find that they possess widely differing colours and would also have a distribution that closely follows the shorelines from whence they originated. When we observe the maria, however, this is not what we find at all. Though it is true that here and there we observe gentle swellings over these vast areas, the deep clefts that are a feature of the maria are of the type associated with mountainous regions and ridges rather than those we would expect from the drying up of an ocean bed,

Photographs taken with large instruments using high magnification do show that there are areas having slightly different shades of colour, but these regions do not follow the contours of the surrounding mountain masses as they would if they had been deposited by tidal currents. As we have seen earlier, such areas can be explained satisfactorily by the igneous theory, as formed by the action of internal forces producing lines of weakness across the maria.

If there has ever been water on the Moon we are, of course, faced with the problem of explaining its disappearance. Either it percolated into the surface and, since measurements made by the radio-astronomers indicate a temperature in the region of $-23\,^{\circ}C$ at a depth of about 2 metres, is still present there in the frozen form, or it evaporated and was lost to the Moon. The answer to the problem clearly hinges on the nature of the lunar surface, particularly that of the maria. Some recent spectroscopic work by Kozyrev, who has examined the reflected light from certain areas of the lunar surface, suggests that quartz is present in fairly large amounts, confirming other work that has indicated a high silica content.

It must be remembered, however, that these studies have covered only a relatively small area, and any conclusions drawn from them may not be generally applicable. This is especially so when we recall that, unless the Moon contains no iron in its core, the mean density of the outer crust must be very low, certainly not higher than about 2·5 times that of water, and only permeable materials like chalk have densities as low as this. Such an observation is, of course, in direct conflict with the work of Kozyrev and other astronomers, and with the preliminary examination of the rock samples brought back from the Moon, which indicates that they may be basaltic in nature. Nevertheless, we cannot definitely rule out the presence of ice below the lunar surface.

On the question of a lunar atmosphere we are perhaps on more certain ground. There are several lines of observational evidence that lead us to the conclusion that the Moon has no atmosphere to speak of; whether it had one in the past is still a matter of controversy. Before making any dogmatic statements on this question, let us first take a look at the evidence available against any lunar atmosphere.

At times, the Moon eclipses the Sun or occults a star, and whenever this happens we have an excellent opportunity of testing for the presence of a lunar atmosphere, since even a very thin one will produce either an aureole around the Moon during a solar eclipse or a gradual dimming of the light of a star during an occultation. In neither case have these been observed. Second, even with very high magnifications, the smallest details on the Moon are seen with perfect clarity; there is no obscuration such as we might expect if they were being viewed through an atmosphere of any appreciable density. Such blurring of detail has, for example, been seen on several occasions during observation of the Martian surface. Lastly, the spectroscope reveals that the light reflected from the lunar surface has a spectrum identical with that of sunlight, there being no measurable changes such as would be found if an atmosphere were present.

Now all of this merely tells us that any atmosphere around the Moon is quite insufficient in density to be detected by any of these methods; it does not provide conclusive proof that no atmosphere at all exists, but indicates only that if it does, it must be extremely tenuous. One method of detecting any trace of atmosphere lies in a refinement of the occultation method mentioned above. From accurate determinations of the lunar diameter and a knowledge of its motion, we can predict the period that should elapse between the disappearance of a star and its subsequent reappearance at the opposite limb. If there is any atmosphere at all it will refract the light and this period will be slightly longer than calculated.

At first sight it might appear that with present-day equipment it should be possible to time the period of occultation sufficiently accurately to distinguish even the smallest trace of an atmosphere if it exists. Unfortunately, there are several difficulties which are not easy to eliminate. The Moon is not a perfect sphere, and the effect of libration – irregularities in its axial motion, which result in our being able to see slightly more than one hemisphere – and

D

49

distortions of the limb due to the craterous nature of the surface combine to make accurate timing extremely difficult. In addition, we are viewing an exceptionally bright, large object against a black background, which also introduces problems of measurement of the lunar diameter. In spite of these difficulties, measurements have been made of a large number of occultations, and the mean values obtained appear to indicate that the period of occultation is slightly longer than that calculated. It is now generally accepted that, although to all intents and purposes the Moon possesses no atmosphere, there may be one having a density about 1/10,000 that of our own.

The Lunar Mascons

So far we have been treating the crust of the Moon as being of essentially uniform density over the whole of the surface. Until a year or so ago, there seemed little reason for astronomers to doubt this assumption. Not until accurate observations of the motions of orbiting spacecraft began to accumulate did it become obvious that there were certain peculiarities in their orbits as they swung around the Moon, irregularities that could be explained only by the presence of masses of dense material lying beneath some of the lunar features.

All of these mascons, as they have been named, are associated with mare regions, and Muller and Sjogren have suggested that they may be asteroids that fell on the Moon with low velocities and are buried below the surface. More recent studies by Kane have shown that not only are the mascons concentrated on the side of the Moon facing the Earth, but they also appear to be closely associated with sinuous rilles, suggesting that they have an igneous origin. If the mascons were produced by impact, they should be randomly distributed over the lunar surface, unless we assume they are due to an asteroid swarm that struck the Moon almost simultaneously. Such a suggestion appears improbable since we have already seen that the maria have widely different ages, as shown by the number of craters that dot their surfaces and the widely differing degrees of erosion of the maria features. Kane believes that the mascons are huge subsurface chambers filled with magma, which would naturally have a higher density than the surrounding material from which the Moon was originally formed. The idea put forward by Gilvarry, that they may be lake sediments that flowed down from the lunar highlands, does not appear likely since we would expect these to have a lower density than the mare material. Whatever the mascons may be, it is possible that we shall learn more about them from further Apollo landings in these regions.

Composition of the Lunar Rocks

The analysis of the lunar rocks brought back by the Apollo 11 mission from the Mare Tranquillitatis has shown that they do not resemble any known terrestrial rocks. Sodium, potassium and rubidium are present in only small concentrations, whereas the titanium, zirconium, chromium and yttrium content is more than twice as high as in any known terrestrial material. Many of the rocks have a crystalline structure and, although it is not possible to say whether they were formed by volcanic action as we know it or by vulcanism triggered off by the impact of meteoritic material, it does seem reasonably certain that they are surface lavas of some form.

The time when the rocks crystallized into their present form has been estimated at between 3,000 and 4,000 million years ago by measuring the amount of radio-active potassium that has been transmuted into argon (an inert gas). There are, of course, some uncertainties associated with this method, but it appears that these rocks are definitely older than the oldest-known rocks on Earth. Further examination of the effect that cosmic rays have had on these rocks has shown that they must have lain close to the surface, not more than one metre down, for at least 20 million years and possibly for as long as 160 million years, attesting to the stability of the lunar surface in this region. Although there are several ray systems crossing the Mare Tranquillitatis, none of these lie in the region visited by the astronauts, and consequently we have none of the ray material available for study at the present time.

As far as the rocks themselves are concerned, two types have so far been identified. The first are known as breccias and appear to be material that has been ejected by the impact of bodies striking the lunar surface. They apparently contain several different types of rock and show minute fractures and also evidence of vitrification, both features being consistent with such a history. Those in the second category are clearly volcanic rocks which were initially flows of lava that has been broken up by impact; they contain many minerals that are well known on Earth, as well as gas cavities.

One common feature of all the rock samples obtained is that their surfaces have been extensively pitted, possibly by micrometeorites, the small indentations being lined with glass. Other glassy beads have been discovered which are of a wide variety of colours, some of them being teardrop-shaped, indicating that they encountered some kind of resistance during their formation. Whether these glassy particles have anything in common with the peculiar fragments known as tektites that are found in large quantities in certain localities on Earth is a point that still requires clarification.

51

The amount of organic material found in the present samples is extremely low, probably less than one part per million, and no hydrated minerals have been discovered, proving that there has been no water on the surface at the landing site during the whole of the period the rocks have been exposed. Everything, therefore, seems to point to the Moon's being a totally dead world. Some atronomers, however, have interpreted certain features of the lunar surface as being due to the presence of some form of life, and we ought to examine the evidence a little more closely.

Life on the Moon

For many centuries, the question of life existing on the Moon in some form or other has posed an intriguing problem for astronomers and biologists, but until comparatively recently one could only speculate on this question. With the rapid development of new techniques and instruments, we can now put forward reasoned and authoritative answers to many of the points raised. The fact that two of the essential ingredients for life as we know it – water and an atmosphere – are lacking on the Moon precludes the existence of any high form of life. Furthermore, the absence of any appreciable atmosphere means that highly lethal radiation from the Sun will fall directly on the surface with no diminution in its intensity, and the fact that the Moon possesses scarcely any magnetic field means that during periods of intense solar activity other charged particles that have a disastrous effect upon living matter bombard the surface without any deflection at all. Cosmic rays also reach the lunar surface from interplanetary space, and all of these factors must be taken into account. We may therefore be certain that, as far as the exposed surface is concerned, no life at all as we know it can possibly exist.

Can we then postulate that there may be bacterial or low plant life in the deep crevices and sheltered caves protected from this incoming radiation? We know, for example, that certain species of terrestrial bacteria can exist in the absence of oxygen and also in arid regions where there is little, if any, water. The fact that no living matter was discovered by the recent Apollo 11 party may be taken as evidence that, on the surface at least, the Moon is completely sterile. The final answer will doubtless come when we are able to study material from the deep crevices and sheltered regions on the Moon.

3. Recent Discoveries in the Solar System

The rapid and spectacular advances in rocketry which, as we have already seen, have led to a tremendous increase in our knowledge of conditions on the lunar surface, have similarly yielded vital information about the two nearer planets – Venus and Mars – and without doubt the next decade will extend this knowledge even further. At the present time, serious consideration is being given to the sending out of unmanned probes into the depths of the solar system to obtain close-up pictures of the outer planets, possibly even as far afield as distant Pluto.

However, we must not overlook the less dramatic, but equally valuable, discoveries that have been made from Earthbound observatories in recent years, nor the various theories which have been put forward to explain the origin of the planetary system. Before discussing this last question, we must first look at some of the general features of the planets since, as always, observation comes first and theories later.

The Terrestrial Planets

The four planets lying closest to the Sun – Mercury, Venus, Earth and Mars – are all very similar in composition. To these we may possibly add Pluto, the most distant planet, discovered in 1930 by Clyde Tombaugh. Since it is so distant and has a very small apparent disc, we know very little about Pluto and, as we shall see later, some astronomers believe that it is not one of the original planets of the solar system, but either a satellite of Neptune which somehow managed to break the chains of gravity and take up a planetary orbit of its own or an interloper from interstellar space that was drawn into the gravitational net of the Sun.

When we examine the mean densities of the first four planets we find that, if we make due allowance for the compression of the material (which is different in each case, depending largely upon the mass of the planet), there is a progressive decrease in average density from about 4·7 for Mercury to 3·9 for Mars, that is, as we move further from the Sun (Table 2).

Table 2

Masses and Mean Densities of the Terrestrial Planets

Planet	Mass (Earth = 1)	Density allowing for compression (Water = 1)
Mercury	0·054	4·7
Venus	0·814	4·4
Earth	1·000	4·4
Mars	0·108	3·9

If we assume that this gradual decrease in density is due to the change in the relative amounts of iron and rock present in these bodies, we arrive at a very interesting situation. The composition of Mercury would be one containing approximately 40 per cent of iron, Venus and Earth having about 30 per cent, and Mars only 20 per cent. When we come to discuss the possible origin of the solar system we shall return to this very important point. In the meantime it is interesting to consider the means by which the iron cores were formed in the case of Mercury and Mars. Now since Venus and Earth are almost identical as far as mean density goes, it is probable that the iron cores were formed in a similar way, and the most likely means appears to be a combination of two heating effects. First, there is undoubtedly internal heating due to radio-activity, this coming predominantly from radio-active potassium and not from uranium as might be expected. Calculation has shown that such radio-active materials will raise the internal temperature to something in the region of 1,500°C. This is well below the melting point of iron which, under conditions like those in the centre of the Earth, is around 5,000°C. This second cause of internal heating is the actual compression of the material of which the planet is composed; although some astronomers have professed doubts as to whether compression alone is sufficient to raise the temperature from 1,500°C to 5,000°C, others such as Hoyle and Gold believe that this may be so, the liquid iron (actually an alloy of iron and nickel with the former being the major constituent) filtering down towards the centre through pores and channels in the rocky mantle.

Now what do we find in the case of Mercury and Mars? Owing to their much smaller masses, compressional heating will have been on a smaller scale. Even so, we must also take into consideration the fact that because of the considerably lower pressures prevailing in

54

the interiors of these planets the melting point of the iron–nickel alloy will be lower than is the case for Venus and Earth, probably not much higher than 1,700°C, which may still lie within the range of heating produced by radio-activity alone. A lot, of course, depends upon the concentration of such radio-active materials in Mercury and Mars, a question to which no definite answer can be given at present. We cannot even be absolutely certain that these two smaller worlds possess high-density cores, although this seems probable.

Mercury, the nearest planet to the Sun, moves in a comparatively eccentric orbit. At perihelion it approaches to within 28,750,000 miles, and at aphelion it swings out to 43,750,000 miles; it makes one complete revolution around the Sun in 87·97 days. Owing to the high inclination of its orbit to the ecliptic, transits of Mercury across the Sun are comparatively rare events, although not so rare as those of Venus. There was a transit on May 9, 1970 and there will be four more transits during the present century: November 10, 1973, November 13, 1986, November 6, 1993, and November 15, 1999.

Whenever a transit occurs, it is noticed that during the ingress and egress there is seldom any sign of a definite bright aureole around the planet as there is in the case of Venus, although several observers have reported one, sometimes bright and sometimes dark, when Mercury is actually in transit across the face of the Sun. This, however, is purely an optical effect due to contrast between the dark body of the planet and the intensely bright solar disc. Confirmation that Mercury has no atmosphere to speak of comes from telescopic observation of the planetary surface.

Several hazy markings have been seen on the planet that are clearly permanent features, and from a long study of their apparent movement Schröter estimated the day on Mercury to be 24 hours 50 seconds in length, very similar to our own. In 1891, however, Schiaparelli concluded that the axial rotation was 87·97 days, equal to the orbital period, which would mean that the planet turns one face permanently towards the Sun. Now owing to the high eccentricity of the orbit, Mercury shows a pronounced libration, alternating in longitude by 23° 7′ on either side of the mean position, one result of this being a gradual change in position of the surface markings with respect to the terminator. Such an effect undoubtedly contributed to the erroneous value found by Schröter for the period of axial rotation.

The ingenious techniques of radio and radar astronomy have now shown that the 87·97-day period is also wrong. Astronomers at the Radio Observatory at Arecibo in Puerto Rico have employed radar techniques to determine the axial period of Mercury and their finding that the planet rotates on its axis in 58·6 days has received additional confirmation both from optical observations and from estimates of the temperature of the dark side, which is somewhat higher than would be expected if the planet kept one hemisphere perpetually away from the Sun.

The old idea that Mercury is locked in to the Sun in the same way that the Moon is locked in to the Earth has had to be abandoned, although not completely so. Colombo and Shapiro have shown mathematically that there is an equilibrium set up between the gravitational forces due to the Sun and tidal effects on the planet which provides a figure in excellent agreement with the observed 58·6-day period for Mercury. Now this has an interesting consequence. The planet is not continuously in the same attitude to the Sun as the Moon is to the Earth but nevertheless there is, in a sense, a very strict relationship between its axial and orbital periods. This is because the 58·6-day period is almost exactly two-thirds of the 87·97-day orbital period. Consequently, as it moves around the Sun, Mercury turns one hemisphere directly to the Sun at one perihelion and the other face at the next. In other words, it makes one complete 'about-turn' during each orbit. This brings about the curious situation that one year on the planet lasts exactly one and a half Mercurian days!

The Mercurian Atmosphere

There are several lines of evidence which indicate that, like the Moon, Mercury possesses no appreciable atmosphere. For example, photometric investigations of the reflectivity of the planet show that it has an albedo almost identical with that of the Moon, and the spectroscope has yielded negative results for the presence of any atmospheric gases. At times, some observers have reported the veiling of certain of the surface features which has occasionally been interpreted as due to atmospheric disturbances, but it appears far more likely that it is caused by suspensions of dust that partially obscure portions of the surface. In the event that an extremely tenuous atmosphere does exist, such very fine particles could remain suspended above the surface for quite considerable periods.

From theoretical considerations alone it is clear that any atmosphere which Mercury may possess must be exceptionally thin. Owing to its relatively small mass, any particles or atoms that travel in excess of 2·6 miles per second will escape completely from the planet (this may be compared with the 7·0 miles per second which is the escape velocity for the Earth and the 1·5 miles per second for the Moon). On the basis of the 87·97-day rotation period, it was thought that the absence of a Mercurian atmosphere could be explained by assuming that the extremely low temperatures prevailing on the dark side of the planet would result in a freezing out of any atmospheric gases with the exception of hydrogen and helium which would, in any event, escape into space in a very short time. Since we now know that every part of the surface of the planet is at some time turned towards the Sun, it is clear that the very high temperatures which will prevail during this period are sufficient for all gases to attain velocities greater than 2·6 miles per second.

Surface Features of Mercury

Because of its proximity to the Sun, it is not easy to observe Mercury although, provided great care is taken, it is possible to observe the planet telescopically in daylight. In general, the planet appears yellowish in colour with ill-defined greyish patches on the surface. One noticeable feature is that the brightness of the disc decreases fairly rapidly near the terminator. There are two probable reasons for this: either the nature of the ground is such that its reflectivity diminishes appreciably whenever the incident sunlight strikes it below a certain angle, or there are differences in the general level which produce irregularities in the form of long shadows near the terminator such as are seen on the lunar surface. Of the two, the latter appears the more probable and the general appearance of the planet shows a striking similarity to that of the Moon, although no detail can be seen owing to its much greater distance. From what we now know of Mars, it seems very likely that the Mercurian landscape is cratered, the greyish patches being reminiscent of the lunar maria.

The Surface of Venus

Moving further from the Sun, we come next to Venus. Although Venus is the most Earth-like of all the other planets and also the closest to us, we know extremely little about its surface conditions. The orbit of Venus is the most circular in the solar system, with an eccentricity of only 0·007, so that its distance from the Sun remains virtually constant at 67,200,000 miles. The inclination of the orbit to the ecliptic is, however, quite high at 3° 23′ 39″ and accordingly transits are extremely rare events, occurring in pairs separated by about eight years. The next transits of the planet will come on June 8, 2004 and June 6, 2012.

Certain markings have been observed on Venus, but these are all of a vague and transitory nature and were recognized as atmospheric phenomena as long ago as the beginning of the nineteenth century, by Herschel and Schröter. Though other astronomers have produced

57

charts of the planet claiming to show more definite markings, which they believed to be true surface features seen through turbulence in the atmosphere, on the whole it appears far more likely that these too were merely of atmospheric origin.

The Rotation of Venus

The length of the day on Venus has been a major problem for many decades. Early studies were all based upon the elusive shadings on the planet and, since these are not now regarded as surface features, such estimates are little more than mere speculation. A value of 23 hours 21 minutes was given by G. D. Cassini in 1666 and similar determinations were made by his son, J. J. Cassini, Schröter and de Vico. By analogy with Mercury, Schiaparelli suggested in 1890 that Venus turns one face perpetually to the Sun and has a rotational period of 224 days, equal to the orbital period. All of these estimates were based upon the assumption that Venus has an axial tilt similar to that of the Earth, a view which is supported by the fact that the cusps of the planet normally appear brighter than the rest of the disc. It must be admitted, however, that there was then very little definite knowledge about the tilt of the planet's axis.

In 1900 Belopolsky attempted to determine the rotational period by means of the Doppler shift in the Cytherean spectrum, and in 1911 amended his original estimate of 24 hours 42 minutes to one of 35 hours. Subsequent work by other observers, particularly by Lowell, supported the much longer period suggested by Schiaparelli. The use of photography to settle the question has generally proved disappointing. Photographs taken in either ordinary or infra-red light reveal scarcely any detail, but in 1927 Ross deduced a rotation period of about thirty days from photographs taken in ultra-violet light which showed certain fairly definite markings that did not appear to arise in the atmosphere.

The detection of radio waves from the planet was announced in 1956 by Kraus, who concluded that they are electrical in origin, possibly similar to those observed from Jupiter, and from certain fluctuations in their intensity he estimated that the day on Venus is one of 22 hours 17 minutes. Although an error of only 10 minutes on either side of this mean value has been claimed, this is not in agreement with the most recent radar studies which show the rotation to be retrograde with a period of approximately 250 days.

Information from the Venus Probes

Now that several close approaches to Venus and soft landings have been made, it is true to say that we have obtained more detailed information about our nearest planetary neighbour than during the entire previous history of astronomy. During 1968, the American space probe Mariner 5 made a close fly-by of the planet, and the Russian probe Venus 4 made a soft landing, telemetering back information on its height above the surface, the composition of the atmosphere, the temperature, and the pressure at various levels. As was previously known from spectroscopic examination, the atmosphere consists largely of carbon dioxide, this gas making up 90 per cent of the total. Curiously, no nitrogen at all seems to exist in the Cytherean atmosphere, while oxygen and water vapour, although present, do not amount to more than 1·5 per cent. From calculations made by Soviet space scientists, it appears that Venus 4 landed in a mountainous region some six miles above what we may, for want of a better name, call sea-level, after a fifteen-mile descent through the atmosphere. The temperature during the descent increased from 40 to 280°C and the maximum pressure recorded was twenty-two atmospheres. These figures are somewhat lower than those obtained by the method of radio-astronomy, which indicates a surface temperature of 500°C and a pressure of close to a hundred atmospheres. Information from Mariner 5 tends to confirm these latter figures. We must not lose sight of the fact that radio-astronomy is a subject in which American scientists excel, and it is possible that the instruments on board Venus 4 were affected in some way by the descent through the thick atmosphere, particularly since the signals from the Soviet probe ceased soon after contact with the surface was made.

Vinogradov has suggested that since the temperature is above 50°C, which is the evaporation point of water, any surface water that may once have been present evaporated and formed a cloud layer which then produced a 'greenhouse' effect. This would be an irreversible process which would raise the temperature to the very high levels recorded. In addition, above 250°C, calcium and magnesium carbonates, which are possibly present in the planet's crust, would decompose with the liberation of carbon dioxide accounting, in part, for the preponderence of this gas in the atmosphere.

All of this information clearly shows how different the Cytherean atmosphere is from our own, but it must be remembered that, although the total percentage of free oxygen on Venus is low, the amount present is not very different from that within the terrestrial atmosphere if we make due allowance for the much higher pressures prevailing on Venus. The major difference would then appear to be the substitution of carbon dioxide for nitrogen on Venus. It is interesting at this point to note that Japanese astronomers have shown that, in spite of this high proportion of carbon dioxide, the atmosphere of Venus will support combustion.

Within the last few years, American astronomers have used radar in an attempt to examine the surface of Venus, and the evidence gathered so far indicates the presence of at least three fairly large areas that stand out above the general surface level and are believed to be mountainous regions. It is quite possible that as these techniques become more refined some sort of map of the planet may be built up similar to those we now have for the Moon, Mercury, and Mars.

The Martian Surface

Beyond Earth lies Mars, a smaller planet than Venus or Earth with a diameter of only 4,200 miles compared with the terrestrial diameter of 7,926 miles. Owing to the quite high eccentricity of the Martian orbit (0·093), its distance from the Sun undergoes considerable variations, ranging from 128,750,000 miles at perihelion to 155 million miles at aphelion; this has a pronounced effect upon the size of the Martian disc when at opposition. The Martian axis is tilted at about 24° and because of this tilt the Martian south pole is exposed at perihelic opposition and the north pole at aphelic opposition. Since the apparent size of the disc is greater when at perihelion, we are able to examine the south polar regions under more favourable conditions and therefore in greater detail.

Even small telescopes will show some detail on the Martian surface, and certain major features have been known and named since the beginning of real telescopic observation of the planet. When it comes to the more delicate markings on the surface, observers vary quite appreciably in what they see; nowhere is this more obvious than in the conflicting descriptions of the tracery of fine lines which has excited more attention than any other feature of this planet. These lines were first described by Schiaparelli during the opposition of 1877 and it is perhaps unfortunate that some astronomers placed too literal a meaning on the name *canali*, meaning literally 'channels', that he gave them.

Percival Lowell, in particular, devoted much of his life to the study of Mars after founding his observatory at Flagstaff in Arizona, in 1894. Whatever their true nature may be, the 'canals' appear to occupy certain well-defined positions on the planetary disc and to be associated with some of the permanent features. Lowell and his colleagues have asserted that certain of them show the phenomenon known as 'doubling', in which a single canal becomes double. It is doubtful, however, if their strict regularity, as shown on many drawings, is real. Many observers have concluded that they consist of small, disconnected regions which, at the distance of Mars, merely seem to be straight lines. This contention has been borne out by the excellent series of pictures telemetered back to Earth by Mariners 6 and 7 which have now provided us with a detailed view of the Martian surface (see Plates V and VI).

One somewhat unexpected result found by the Mariner probes is that the surface of Mars is very similar to that of the Moon, being mainly craterous in appearance. There are, however, certain differences in the Martian craters. Since Mars possesses an atmosphere, the walls of the craters show quite distinct signs of erosion and are not as cleanly sculptured as the corresponding lunar features. Several of the craters, too, are arranged in almost perfectly straight lines, which may account, to a certain extent, for the appearance of canals. Tolansky has suggested that Mars may have been formed initially very close to the Earth and at the same time as the Moon, the planet being projected into its present orbit during the early period following the formation of the solar system. Since the Martian atmosphere is so tenuous, it is perhaps not too surprising that the surface is craterous since the larger meteorites, which presumably formed some of the lunar craters, would also reach the Martian surface before burning up during their descent.

The Martian Atmosphere

The Martian probes have also provided us with detailed information on the constitution of the Martian atmosphere. Here again we encounter the curious fact that no nitrogen appears to be present on Mars, much of the atmosphere consisting of carbon dioxide, with some free oxygen and water vapour also present. The situation is therefore analogous to that found recently on Venus, although the Martian atmosphere is much less dense. As yet, the composition of the polar caps is still in doubt. They may consist either of solid carbon dioxide or of a thin layer of ice. One very important observation is that in the region of the south polar cap both ammonia and methane have been detected – two of the constituents necessary for the beginning of life – and if liquid water does exist it is possible that certain very primitive life forms may also have come into existence on Mars. Should this indeed be the case, it is extremely important that no contamination from Earth be introduced by any manned or unmanned landings, since this could quite easily invalidate any conclusions reached on the existence of indigenous Martian life.

The existence of an atmosphere on Mars has been known for some time and, on occasion, veiling of some of the more prominent features has been observed; this can be most satisfactorily accounted for on the basis of dust storms carried across the surface of the planet by wind currents. One further aspect of the Martian atmosphere deserves mention here. When photographs of the planet are taken in blue light, practically no trace at all of the major surface markings can be seen; in other words, the atmosphere is virtually opaque to blue light, far more so than our own atmosphere. During the last two weeks of May 1937, however, as a result of an abrupt and dramatic change in the Martian atmosphere, photographs taken in blue light at the Lowell Observatory showed not only the Syrtis Major but also many other features with unprecedented clarity. That this change was atmospheric in character was proved by simultaneous photographs taken in red and yellow light, which revealed no difference in the surface markings. So far, no satisfactory explanation has been given for this curious, and temporary, phenomenon.

Readings of the temperature on the Martian surface taken by means of a thermocouple have indicated that whereas at noon on the equator it may be around 15°C, in the polar regions it can be as low as —90°C and, since the atmosphere is so thin, much of the daytime heat will be radiated away very rapidly during the night, resulting in extreme fluctuations in temperature, not only with latitude but also throughout the Martian day and night.

All in all, Mars has been shown to be a somewhat forbidding world and the recent pictures obtained by Mariners 6 and 7 have drastically altered our ideas about the planet. It seems true to say that as further data is accumulated we may have to change them still more.

Mars possesses two tiny satellites – Deimos and Phobos – which, from estimates of their brightness, are only five and ten miles in diameter respectively. At present we know nothing of their composition or the nature of their surfaces, but from small variations in their reflected light it appears certain that they both keep the same face turned towards the planet. This is the situation we would expect on theoretical grounds, since they are both extremely close to the planet, revolving in almost circular orbits, Deimos some 12,500 miles above the Martian surface and Phobos only 3,700 miles.

Pluto

Pluto, the remaining terrestrial planet, is something of a curiosity. Unlike the intermediate giant planets which lie between it and the Sun, it is only slightly larger than Mercury and has the most eccentric of all planetary orbits. At its closest, it is only 2,700 million miles from the Sun, coming within the orbit of Neptune, while at aphelion it moves out to more than 4,600 million miles.

The small mass, estimated at about 0·45 that of the Earth, coupled with this high eccentricity of the orbit, has led many astronomers to believe that Pluto was originally a satellite of Neptune which was permanently lost during the early stages of the formation of the solar system. Why it should have escaped from the gravitational field of Neptune is still something of a mystery. One possible explanation is that radiation pressure from the Sun removed large quantities of gas from the protoplanet that later became Neptune, reducing its mass – and thereby its gravitational pull – sufficiently to allow Pluto to break free. At the same time Triton, another satellite of Neptune, may also have escaped, only to be recaptured during a chance re-encounter.

The surface temperature on Pluto is only about 50°K, not much above absolute zero, and any atmosphere it may originally have possessed will have been frozen upon the surface by now.

The question of further planets lying beyond the orbit of Pluto has been discussed on several occasions during recent years, especially since any trans-Neptunian planet was expected to be similar to the giant planets in constitution and not of a terrestrial nature but, although present instruments are capable of picking up any planetary object of the size of Neptune out to five or six times this distance, nothing has so far been discovered.

There remains, however, the possibility that the diameter of the disc which we measure telescopically is not the true diameter of the planet. It has been suggested that the surface of Pluto acts rather like that of a highly polished ball which concentrates the faint sunlight reaching it into a bright spot and it is this which is actually being measured. In view of recent observations of Pluto, this explanation now appears unlikely.

Some American astronomers are now urging that an unmanned tour of the outer planets should be undertaken in 1977 when a rather rare opportunity arises – one that occurs only every 175 years – when the gravitational pull of the giant planets may be utilized to swing a space probe out towards Neptune. Such a journey would only take about eleven years, as compared with the thirty years or so that would be required by means of ordinary rocket propulsion. If such a mission can be successfully carried out, it will provide much-needed information concerning these outer planets of the solar system and may solve the problem of Pluto once and for all.

Origin of the Inner Terrestrial Planets

Before we leave the question of the terrestrial planets, we must consider a recent and somewhat startling theory regarding the possible common origin of these bodies. The fact that all the planets appear to be revolving in stable orbits about the Sun undoubtedly led to the conclusion, reached by the early astronomers, that the planets have always occupied these paths about the Sun since their formation. Indeed, until recently there was little evidence to the contrary.

In 1960, however, Lyttleton suggested, from a study of the characteristics of rotating fluid bodies in combination with the dynamic properties of the solar system, that there may have been two unstable planets, one of which broke up to form Mercury and Venus, with the other forming the Earth, Moon and Mars. By using the most recent values of the masses and volumes of these bodies, McCrea has shown that the densities of these two protoplanets would have been virtually identical, namely 5·6 times that of water. Such a theory overcomes a very curious anomaly. When we examine the masses and densities of the four inner terrestrial planets, together with that of the Moon, we find that there is no correlation at all between these two parameters. This is clearly seen from the figures given in Table 3.

Table 3

Masses and Mean Densities of Terrestrial Planets

Planet	Mass (10^{24} kg)	Density (Water = 1)
Moon	0·073	3·34
Mercury	0·330	5·47
Mars	0·639	3·91
Venus	4·872	5·24
Earth	5·974	5·51

If we take into account the compression of the material in the two hypothetical protoplanets we find that the densities come out to be equal, indicating that their composition was the same. Now this appears to conflict with the figures in Table 3, for clearly the densities of the resulting bodies are not at all similar; in particular, those of the Moon and Mars are markedly smaller than the others.

How can we explain this wide discrepancy on the present theory? The answer is by assuming, as McCrea has shown, that the break-up of the two protoplanets, which in both cases was due to their rotational instability, occurred at different stages of their development. Once such a protoplanet has formed, segregation of the constituents will take place, with the denser material moving towards the core and the lighter components forming the mantle. If, now, the rotation of such a body was such that it produced instabilities in the core but not throughout the whole of the mass, then upon break-up of the core the larger portion would take with it more of the low-density mantle. Such a break-up is envisaged by McCrea as the origin of Venus and Mercury and would satisfactorily explain the higher density of the latter.

Similarly, if the second protoplanet became rotationally unstable as a whole, we find that the formation of three independent bodies would mean that the smaller ones would have the lower average densities, since they would be formed from the material of the mantle rather than of the core. Thus these two different modes of break-up will explain the observed densities, but what of the rotations? Once again, as Lyttleton has pointed out, we find a difference. Whereas the periods of rotation of Earth and Mars are both of the order of twenty-four hours, those of Venus and Mercury are much longer.

There is, of course, one further problem which we have not so far mentioned. Even the combined masses of the Earth, Mars and the Moon fall far short of that of Jupiter, yet this planet is apparently stable even though rotating extremely rapidly. The answer appears to lie in the widely different chemical constitutions of Jupiter and the protoplanet. The comparatively light materials from which Jupiter was formed, and of which it is now composed, are such that it can rotate quite rapidly without any instabilities arising, whereas the unstable protoplanets we have just been discussing are thought to have been formed from fairly slowly rotating bodies which, although containing quite a high proportion of hydrogen and light gases, became unstable owing to their contraction, which would bring about a large increase in the rate of rotation. Most of these light gases would be lost during subsequent stages of evolution. Later in the chapter we shall be discussing another theory of the formation of planets from smaller, cold bodies and, as we shall see, this method cannot lead to the type of rotational instability we have just been considering.

The Giant Planets

The four major planets which lie beyond the orbit of Mars are quite different in their size and constitution from those we have just been discussing. Their mean densities are so low that there is no question at all of their possessing cores of rock or iron. They are, nevertheless, far more massive than any of the terrestrial planets. The very low densities, as shown in Table 4, have presented astronomers with a problem that is only just being solved, namely, the internal constitution of these planets.

Table 4

Masses and Mean Densities of Major Planets

Planet	Mass (Earth = 1)	Density (Water = 1)
Jupiter	318·4	1·35
Saturn	95·3	0·71
Uranus	14·6	1·56
Neptune	17·3	2·47

Until a few years ago, it was thought that the very low densities could be explained on the assumption that the giant planets contain not more than a maximum of 40 per cent hydrogen in their make-up. From theoretical considerations, however, which will be explained later, we would expect them to contain far more, in fact to be more similar to the Sun in their chemical composition. This figure of 40 per cent hydrogen does not make sufficient allowance for the effects of compression, particularly in the case of Jupiter and Saturn, and more recent recalculations by Harrison Brown and Ramsey have indicated that these two planets must contain as much as 80 per cent of hydrogen.

Jupiter

Jupiter, lying at a mean distance of 482 million miles from the Sun, is more than twice as massive as all of the other planets put together yet, in spite of its great size, it rotates on its axis in only 9 hours 50 minutes. The result of this rapid rotation is that its shape is far from spherical, the polar axis being only $\frac{15}{16}$ of the equatorial diameter.

The Jovian atmosphere, which is many hundreds of miles deep, is composed mainly of hydrogen, ammonia and methane, and it is interesting to note that such an atmosphere could support some very early life-producing activity, being in some ways similar to the primeval terrestrial atmosphere which, as we shall see in a later chapter, was a reducing one containing very similar gases. The present oxidizing atmosphere on Earth is a secondary one produced by a very complex series of chemical and physical changes.

The core of Jupiter is completely hidden beneath the turbulent atmosphere and it is extremely doubtful that it consists of rock or metal. It is far more likely that it is composed of ice or solid hydrogen, compressed by the tremendous gravity and weight of the overlying layers of gas to a metallic consistency. Though it seems scarcely credible, under the conditions prevailing on the Jovian surface solid hydrogen may be as hard as steel, whereas steel itself would be as brittle as glass.

In the telescope, the planet presents a curiously banded appearance, these zones being visible even in moderate-sized telescopes. Only the regions around the poles are of a uniform greyish colour. With more powerful instruments it is possible to see finer detail within the zones (see Plate VII). Dark spots and streaks, often connecting two zones, are of frequent occurrence, together with much brighter spots and, by accurate timing of the passage of these features across the planet's central meridian, it is possible to demonstrate that, like the Sun, the atmosphere does not rotate as a solid body but that each zone has its own particular period of rotation. None of the smaller features have any real permanence; they are obviously local and violent disturbances within the atmosphere. As yet, there is no evidence at all of any periodicity in these small-scale changes.

In the midst of all this change, however, one feature, which is readily visible in small telescopes, has been more or less prominent since its discovery in 1831. This is the Great Red Spot, which lies in a bay just south of the Jovian equator. At times its red colour is extremely noticeable; at others it fades so as to be virtually invisible. There has been considerable argument over the nature of this feature for several decades. Since it shows an appreciable drift in longitude, it is clearly an atmospheric phenomenon and not a surface feature although it is possible that at its lowest point it is in contact with the surface. It may be some sort of internal moon, which has not succeeded in breaking free of the planet's tremendous gravitational pull. Other astronomers have suggested that it may be an island of solid helium floating in the Jovian atmosphere, but this seems unlikely for several reasons. Solid helium is not coloured and, since it is also the coldest substance known, its surroundings must be at a higher temperature; under such conditions solid helium would be thermally unstable and would doubtless melt in a short time. More likely, the Red Spot is a solid compound of sodium and ammonia.

67

A few of these compounds have been prepared in the laboratory and all are strongly coloured.

Towards the beginning of the present century, a second semi-permanent feature known as the South Tropical Disturbance was discovered; it shows an even more pronounced longitudinal drift than the Red Spot, so much so that by 1918 it had extended almost half-way around the disc of the planet. Whenever the South Tropical Disturbance overtakes the Red Spot in its passage along the south tropical zone, violent atmospheric changes take place which often result in radio emission from the planet.

Radio Emission from Jupiter

Sporadic radio emission from Jupiter in the decametre wavelength range has been known for some time and various theories have been advanced to explain it. From the extremely violent motions of the various features within the Jovian atmosphere it is perhaps only logical that it should have been initially explained on the basis of thunderstorm activity, an idea that has been extensively examined, particularly by Shain. That part of the radio emission may be attributed to such atmospheric phenomena, which must be on a far vaster scale than anything experienced on Earth, seems quite plausible. Undoubtedly there are cyclonic winds of incredible force within the deep murk of the Jovian atmosphere. In a similar manner, as Sagan and Miller have suggested, such radio emission may also arise from chemical explosions, a view that is particularly apt since there are several very reactive gases present in the atmosphere existing under extreme conditions of pressure and temperature.

More recently, Strom and Strom have put forward the idea that the magnetosphere surrounding Jupiter may focus radio waves coming in from very distant sources, and the possibility that the ionosphere may oscillate violently, giving rise to decametre radiation, has been examined by Zhelezniakov. Unfortunately, all of these ideas encounter serious theoretical difficulties and it now seems far more probable that the radio waves originate in a Van Allen belt around the planet similar to that discovered around the Earth, particularly since the decametre radiation appears to be closely related to events taking place on the Sun. On this basis, we may look upon the Jovian radio emission as due to electrons from the Sun that are accelerated to extremely high velocities within the Jovian Van Allen belt and are dumped into the magnetic polar regions of the planet. Though a great deal of theoretical work has already been carried out on the problem, much still remains to be done before a complete answer to this very intriguing question can be given.

The Satellites of Jupiter

Of the eleven known satellites of Jupiter, four are very much larger than the rest, and the three outer members all move in a retrograde manner. The four large moons – Io, Europa, Ganymede and Callisto – have the distinction of being the first of the heavenly bodies to have been discovered by means of the telescope. Both Ganymede and Callisto, with diameters of 3,270 and 3,140 miles respectively, are larger than the Moon. Although in large telescopes they appear to show markings similar to those of Jupiter itself, neither they nor any of the other satellites have atmospheres. Judging from the retrograde motions of the three outer satellites which, from their brightness, appear to have diameters less than thirty-five miles, it is possible that these three members may be merely asteroidal bodies captured by Jupiter rather than original satellites of the planet.

Saturn

A planet that is, in many respects, even more extreme than Jupiter is Saturn. Lying at a mean distance of some 888 million miles from the Sun, it takes 29·5 years to complete one orbit and rotates so rapidly on its axis – in a 'day' of only 10 hours 14 minutes – that, like Jupiter, it is appreciably flattened at the poles. Since its axis is inclined at an angle of 26° 44' to the perpendicular to the plane of the orbit, the planet's orientation to the Sun is continually changing and as a result the seasons on Saturn are quite marked. These seasons last about 7·5 years and also mark the changes visible in the unique ring system which lies in the equatorial plane of Saturn (see Plate VIII). Although the rings are extremely wide, about 170 thousand miles in diameter, they are only an estimated ten miles in thickness.

The Ring System of Saturn

Since they are unique in the solar system, it is worthwhile considering the rings of Saturn in some detail. They are seen to their best advantage whenever the planet is at one or other of its solstices, since they are then at their most open phase. As the planet approaches the equinoxes the angle of tilt diminishes rapidly until at the equinoxes themselves the rings are seen edge-on and are quite invisible except in very large instruments.

From Fig. 6 it will be seen that when the rings are situated in the same plane as the Earth and the Sun they disappear twice as seen from the Earth, whereas seen from the Sun they will disappear only once. This is entirely due to the fact that the Earth is moving with respect to both the Sun and the rings, and it is scarcely surprising

69

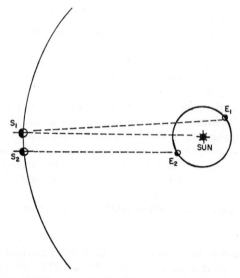

Fig. 6 Diagram showing the double disappearance of the rings of Saturn as seen from the Earth.

that this peculiar behaviour of the rings of Saturn excited the curiosity of astronomers shortly after the invention of the telescope and led to both puzzlement and confusion until their true nature was appreciated.

Broadly speaking, it is possible to distinguish three main rings: an outer ring *A*, which is greyish in colour and is separated from the much brighter ring *B* by Cassini's division, and a darker, inner ring *C*, which has also been named the crepe ring and through which it is possible to see the globe of the planet itself. In large telescopes, using a high magnification, all three rings may be further subdivided into several more, some of which are more prominent than others. Many of these secondary rings, however, appear to be little more than divisions into regions of varying brightness, such regions, unlike the three already mentioned, being only semi-permanent. Several gaps other than Cassini's division also exist within the three main rings, for example, Encke's division, which divides the outer ring *A* into parts of unequal brightness.

This brings us to a consideration of the nature of the rings. At times, Saturn occults a fairly bright star and on such occasions it is possible to observe a dimming of the star as the rings pass in front of it. Such observations have shown that the three main rings do not have the same opacity. Ring C is the most transparent (we have already mentioned that the globe of Saturn is plainly visible through it), ring A is only slightly transparent, and ring B is practically opaque. This opacity of the rings is not compatible with their being either fluid or gaseous in nature. The only explanation that is in agreement with the observed facts is that they are composed of small particles of either rock or ice, free to act independently under the powerful influence of Saturn and possibly certain of the inner Saturnian satellites. The dynamics of the ring system completely rules out the possibility of their being wholly solid. This idea has received confirmation from spectroscopic measurements of the velocities within the different zones of the rings. The outer edge of ring A revolves in 14·5 hours while the inner surface of ring C takes only 4 hours for one revolution.

What remains now to be considered is why Saturn should possess this unique ring system at all. For example, is it the remains of a moon that somehow exploded in past ages, or even of a satellite that did not form properly owing to the tremendous gravitational field of the planet? At present it is not possible to say which of these two possibilities is correct, since either would result in very much the same situation as that which we now find.

What we are able to say, however, is why such a system of rings should be formed instead of a very close satellite. The mathematical theory was first worked out by Roche and depends basically upon what is known as Roche's limit. If we have two bodies of unequal mass revolving around a common centre of gravity, it is found that there exists a critical distance within which they cannot approach each other without serious instabilities arising within the system. This limit depends upon the individual masses of the two components and, provided the two bodies are of the same density, the radius of the region around a planet in which its tide-producing force is stronger than the gravitational force of a satellite (that which will hold it together as a coherent body) is 2·44 times the radius of the planet. Outside this limit, the satellite will not be disrupted, but only disturbed from a spherical shape.

If, therefore, at some time in the past, Saturn possessed a small inner satellite that came within Roche's limit for this planet, it would be broken up into small fragments which would then, under the influence of Saturn's gravity, form into a ring. The much smaller influence of the other satellites produces the gaps that we now find in this ring system. On the other hand, the rings may represent a still-born moon that was unable to form into a solid body since the particles of which it would have been composed lay inside Roche's limit. Quite clearly, if any other planet had had a body rotating sufficiently close to it, such a body would also have been fragmented into a ring system. Far from being contrary to what we might expect, therefore, the rings of Saturn merely prove that the laws of nature are the same everywhere in the universe.

The Constitution of Saturn

Like Jupiter, Saturn presents a banded appearance in the telescope, although the bands are not as noticeable as in the case of Jupiter. Whether this is due to the much greater distance from the Earth or whether these atmospheric features are intrinsically fainter is a question still open to debate. In addition to the bands, which incidentally are somewhat more regular in outline than those visible on the Jovian disc, there are also light and dark spots which may be used to determine the rotation periods within the various Saturnian latitudes. These show that the velocity of rotation decreases from the equator towards the poles and suggest that, as with Jupiter, there is a deep gaseous atmosphere which possibly rests upon a semi-solid layer at much deeper levels. Spectroscopically, we find that the major constituents of the Saturnian atmosphere are hydrogen, methane and ammonia, the latter probably existing in the solid state since the atmospheric temperature is in the region of $-150°C$. This is borne out to a certain extent by the fact that, whereas in the case of Jupiter it is ammonia that predominates, in Saturn methane is the major component.

The Satellites of Saturn

Beyond the rings, and lying in almost the same plane, are the various moons of Saturn, of which nine were known until recently. A tenth satellite, which was first seen late in 1966 by Dollfus, has been named Janus. Like all of the Saturnian satellites except Titan, Janus is quite a small body and invisible in small telescopes. It is the innermost of all the satellites of this planet, lying very close to the outer edge of the ring system and, owing to the glare of Saturn, difficult to see except at its greatest elongation. Since it must lie very close to Roche's limit, it is probable that Janus is not spherical. Quite possibly, future estimates of any variations in its brightness will resolve this point.

Titan, the largest of the satellites, with a diameter of 3,550 miles, is unique in that it has an atmosphere, discovered spectroscopically by Kuiper. Like Saturn, this atmosphere consists mainly of ammonia and methane. The fact that Titan has an atmosphere at once raises some very interesting points. The suggestion is strong that both it and Saturn had a common origin and this must clearly be taken into account in any theory of the origin of the solar system. Not only this, but the mass of Titan is so small that, for any atmosphere to exist, the temperature must always have been extremely low. A rise of only 80°C or so at any time since the satellite's formation would have resulted in this atmosphere's being completely lost, for even this small rise in temperature would boost the velocity of the molecules above the escape velocity of Titan. If, then, all of the bodies that make up the solar system have passed through a period of high temperature, as suggested by certain theories of their formation, it is clear that the atmosphere of Titan must have been formed some time after the satellite itself. In marked contrast to the other planetary satellites, Titan has a pronounced orange colour, which is explained by Kuiper as due to oxidation of the surface by the atmosphere.

The very refined techniques used to prove the presence of an atmosphere around Titan have also been used to search for similar tenuous atmospheres around the larger satellites of Jupiter but, in spite of the fact that these bodies are much closer to us and any atmospheric lines in their spectra should be more readily discernible, no atmospheres have been discovered, even for the largest Jovian satellites, such as Ganymede, Callisto and Io, which have comparable diameters and presumably similar masses and compositions. Why this should be so is still something of a mystery.

One further satellite of Saturn deserves some mention here. The tiny moon Phoebe, with a diameter of only a hundred miles or so, has an inclination to the equatorial plane of the planet of more than 150° and revolves in a retrograde manner. Possibly the same explanation as given for the three outer moons of Jupiter also applies here.

Uranus

We must travel a thousand million miles beyond the orbit of Saturn before reaching the next giant planet of the solar system, Uranus, which revolves around the Sun at a mean distance of 1,783 million miles, its orbit lying almost exactly in the plane of the ecliptic. Since its orbital velocity is very small, amounting to only 4·25 miles per second, the 'year' on Uranus is one of 84 years 4 days and an observer on Earth is unlikely to see one complete revolution of this planet around the ecliptic.

Although Uranus exhibits a discernible disc – which appears of a greenish-blue colour – in small telescopes, large instruments are necessary for satisfactory observation of the planet. Like the two

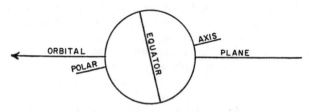

Fig. 7 The peculiar inclination of the axis of rotation of the planet Uranus.

giant planets already discussed, Uranus is flattened at the poles owing to its rapid axial rotation, in a period of about 10 hours 49 minutes. Faint bands are visible on the disc, lying parallel to the equator; they are dark grey in colour and show little discernible detail, this lack being almost certainly due to distance rather than to any intrinsic absence of detail.

The unique feature of Uranus is its axial tilt, the axis of rotation being inclined at an angle of 98° to the perpendicular to the plane of the solar system (Fig. 7). This means that as it revolves in its orbit Uranus is virtually lying on its side, so that when the planet is at its solstices the north or south pole is pointed almost directly towards the Earth and consequently the disc of the planet is then almost circular since the equator is facing us and not the meridian. In addition, the bands will be almost invisible during this period, being greatly foreshortened by perspective.

The spectroscope tells us that the atmosphere of Uranus is composed predominantly of hydrogen and methane, any ammonia that may be present being certainly in the solid state at the low temperature of −160°C which prevails there, and frozen on to the invisible surface.

The Satellites of Uranus

Five satellites are known at present, all lying in the equatorial plane of the planet and having diameters ranging from 200 to 700 miles. Miranda, the smallest of the five, was discovered photographically by Kuiper in 1948. The orbits of the five satellites are almost circular and, since they revolve in the plane of the planet's equator, they appear to move in a retrograde fashion owing to the large axial tilt of Uranus. At the moment very little is known of the constitution of these small bodies.

Neptune

A further large gap separates Uranus from Neptune, the last of the giant planets. In size, Neptune is so like Uranus that first one and then the other has been estimated to be the larger of the two, with different astronomers still quoting different figures for their diameters. The period of axial rotation has been derived from spectroscopic measurements as 15 hours 48 minutes and, although it is not possible to measure the oblateness of the planet directly, the precessional movement observed in the orbit of Triton, its nearer and larger satellite, has proved that Neptune is indeed flattened appreciably at the poles. The mean distance of the planet from the Sun is 2,792 million miles, and with an orbital velocity of only 3·4 miles per second it requires 164 years 289 days to make one complete revolution around the Sun. As a result, Neptune will not return to the position of its discovery until 2011.

The Constitution of Neptune

The spectroscope has revealed that the atmosphere of Neptune is very like that of Uranus, consisting mainly of hydrogen and methane with possibly neon present also. At a temperature of −170°C, the methane is probably mostly in the liquid form and any ammonia present is certainly solid under these extreme conditions. Visually, a banded appearance may be made out under high magnification, although at this distance very little detail at all is apparent.

The Satellites of Neptune

Neptune has two known satellites. The larger, Triton, was discovered by Lassell in 1846; it is somewhat larger than our Moon, with a diameter in excess of 5,000 miles, and it revolves in a retrograde orbit only 220,000 miles from the planet. Owing to the equatorial bulge of Neptune, Triton precesses in about 580 years, the orbit being inclined by 159·9° to the planet's equator.

If we accept the figure of 5,800 miles for the diameter of Triton that has been calculated from its brightness when at mean opposition, then clearly this satellite is intermediate in size between Mars and the Earth. Its mass has been estimated by Alden as 1·80 times that of the Moon.

Nereid, the remaining satellite of Neptune, has a diameter of only about 200 miles and was discovered photographically by Kuiper in 1949. Unlike Triton, it has a highly eccentric orbit (Fig. 8) and its motion about the planet is direct.

The Minor Planets and Meteorites

The minor planets, or asteroids, were unknown for almost two centuries after the invention of the telescope, not because they can

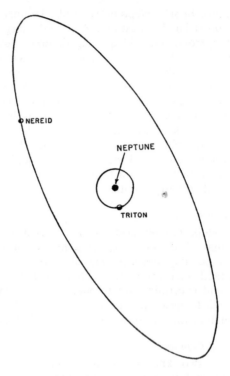

Fig. 8 The orbits of Neptune's satellites.

be seen only in large instruments – several are readily visible in even small telescopes – but because they are so small that none show an appreciable disc and it is possible to pick them out from the myriad of background stars only by their motion.

The seventeenth-century astronomers were quite aware that the gap between Mars and Jupiter was wider than would be anticipated from a study of the orbits of the known planets and several suggested that an undiscovered planet should lie beyond the orbit of Mars. The first such body was found by Piazzi on January 1, 1801 and named Ceres, but from its brightness alone it was quite obvious that this was not a planet in the accepted sense of the word. The following year Pallas was discovered by Olbers, and Harding found Juno in 1804; a fourth asteroid, Vesta, was discovered by Olbers in 1807. When at a favourable opposition, Vesta is just visible to the naked eye on a clear, moonless night.

76

The fact that four planetoids had been found between Mars and Jupiter in place of the single large planet that had been expected naturally prompted further searches, but it was not until 1845 that any more members of this family were found; from 1847 onward more asteroids were added to the list every year. In the ten years from 1852 to 1861, for example, Goldsmith alone discovered no fewer than fourteen minor planets, using only a small refractor, and by 1890 more than 300 were known.

The Discovery of the Asteroids

Initially, the method used for discovering these tiny bodies was to examine the sky in the region of the ecliptic telescopically and compare the star fields with the charts available at that time. If a new star was found it would be closely watched for several nights in order to detect any motion, which would indicate that the star in question was in reality an asteroid. In 1891, Max Wolf introduced the photographic method, which has been used ever since. Being members of the solar system, the asteroids move against the background of stellar points and consequently if, during the exposure, the telscope is driven at such a rate as to follow the stars exactly, any asteroids present in the field will show up as short trails on the photographic plate owing to their differential motion. A second method is to examine two photographs taken a few days apart, by means of a stereoscope, when the asteroids will appear to stand out against the star fields which may be reckoned to be at infinity. By this means very faint asteroids can be picked out against rich star fields.

At present a little over 2,000 of these tiny bodies are known and have been catalogued. One of the major difficulties is keeping track of them all, since to do this the orbits must be calculated from the available data. There have been several instances of an asteroid being lost and then rediscovered as a different body, and a great deal of involved mathematical work is necessary in order to prove its identity.

Dimensions of the Asteroids

Only about a score of these bodies have diameters in excess of 100 miles and even these are extremely difficult to measure accurately. The most recent values for the largest are given in Table 5.

The diameters of the smaller ones have been estimated in a rather indirect manner by assuming that their surfaces are similar in nature to that of Mercury or Mars and that the amount of sunlight they

Table 5

Diameters of the Largest Asteroids

Asteroid	Diameter (miles)
Ceres	437
Pallas	287
Vesta	240
Hygeia	200
Eunomia	179
Psyche	175
Davida	162
Juno	140
Hebe	138
Iris	125

reflect is proportional to their surface areas. On this basis, it has been reckoned that there may be around 30,000 that are a mile or so in diameter and many millions that are no more than boulders drifting through interplanetary space.

The large majority of these bodies are far from being spherical in shape, although the four largest would appear to be so. From the variations in their brightness, which have been explained as due to the presentation of different portions of their surfaces to the Earth, it has been concluded that most of the minor planets are far from regular in outline, and one in particular – Eros – exhibits such curious changes in brilliance that several different hypotheses have been advanced to explain them. For example, it may be a very elongated piece of rock that is rotating about its minor axis, or even a pair of very small bodies bound together by gravity, or it may have a somewhat crystalline structure with the various facets reflecting light to different degrees. Of these, the first would appear to be the most plausible, but further observational data are necessary before the question can be finally settled.

Distribution of the Asteroids

Most of these bodies have orbits that lie between Mars and Jupiter, but the general picture of their distribution within the solar system is much more complex and there are several that move in highly elliptical orbits, some coming in quite close to the Earth and at least one – Icarus – approaching the Sun within the orbit of

Mercury. During June 1968 Icarus made a very close approach to the Earth and was well placed for observation, since at its nearest it was only 4 million miles away. Since Icarus has an orbit extremely like that of a typical comet, there has always been the possibility that it might not be an asteroid at all but a comet with a bright nucleus and no coma. Both visual and photographic observations have shown quite conclusively that this is not the case. Indeed, observation of its variations in brightness have yielded a period of about two hours for its rotation upon its axis. Photometric investigations have also shown that it has a bluer colour than the normal asteroids.

On June 13, 1968 the MIT Laboratory at Lexington, Massachusetts, made radar contact with Icarus and nineteen hours later the Jet Propulsion Laboratory at Goldstone, California, also picked up the asteroid, radar observations continuing until June 16. When the echoes that were received back from this body have been fully processed, it is confidently expected that they will yield extremely valuable information concerning the nature of the surface of this tiny planetoid. In addition, the orbit of the asteroid has now been calculated with extreme accuracy, and since Icarus passed within 10 million miles of Mercury on May 1, 1968 this will provide vital information on the theory of relativity, particularly on whether natural events are better described by the simple tensor relativity of Einstein or the more sophisticated scalar-tensor mechanics of Dicke.

The asteroids Apollo and Adonis both approach the Sun closer than the orbit of Earth and at times are closer than Venus. In 1937 the tiny asteroid Hermes was discovered by Reinmuth and in that year it passed within 400,000 miles of the Earth. The orbit of this minor planet is the smallest of all, crossing that of the Earth almost exactly. Owing to this coincidence with our own orbit about the Sun, Hermes can come with 221,000 miles of the Earth, closer even than the Moon. Unfortunately, there are still some uncertainties associated with the orbit of Hermes and the foregoing figure may eventually have to be modified somewhat.

This naturally raises the question of the effect on the Earth of a direct collision with a body such as Hermes. The chance of such a collision is, fortunately, extremely rare. Nevertheless, the Earth has been struck by bodies of this size in the past and various large craters exist in different parts of the world to testify to the catastrophic force of such impacts. It may be of interest here to examine some of the differences between such craters as found on the Earth and those that exist on the Moon for, as we saw in the last chapter,

79

some at least of the lunar craters must have been formed by similar meteoritic impacts upon the lunar surface.

The main difference, of course, will be due to atmospheric effects. Not only will friction with the atmosphere of the Earth heat the outer surface of a meteorite to incandescence but there will also be a pressure wave travelling at a high velocity ahead of the falling body. This will have the effect of reducing the forward velocity of the meteorite to a certain extent, so that the velocity of impact will be lower than in similar circumstances on the Moon. We cannot, however, neglect the influence of this pressure wave upon the Earth at the point of impact, for calculation shows that this itself will produce a tremendous shattering effect on the surface. Once a terrestrial meteorite crater has formed, erosion too plays a major part in its subsequent shaping, a process which is virtually absent on the Moon.

Meteorites are also important, not only to the astronomer, but to the biologist, since they are the only extra-terrestrial bodies to reach us and have yielded extremely important information on the possibility of life existing elsewhere in the solar system. This, however, is a complex problem and one that will be deferred until Chapter 14.

Comets

One final class of bodies belonging to the solar system that have not been mentioned so far are the comets. Although they appear in any part of the heavens and are by no means confined to the region along the ecliptic, they still obey the same laws as the planets and satellites. It is simply because their orbits are inclined at considerable angles to the plane of the solar system and their positions are constantly changing with respect to the Earth that cometary motions appear to be so different from those of the planets.

Cometary Orbits

So far as is known, no comet revolves about the sun in anything like a circular orbit. Almost all their orbits are either very elongated ellipses or parabolas; one or two are slightly hyperbolic. Now what determines whether a comet moves in an ellipse, a parabola or a hyperbola? The answer is its velocity. Depending upon its distance from the Sun, there is a certain critical speed known as the circular velocity, with which an object will move in a circle. If the body moves at a slightly higher velocity, its path will be in the form of an ellipse and the ellipse becomes more eccentric as the velocity becomes greater. However, if the speed of the object is 1·414 times

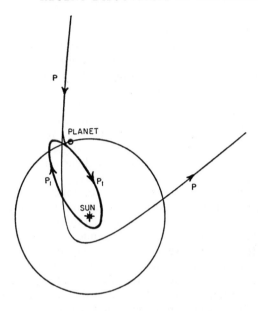

Fig. 9 Diagram showing the change of a cometary orbit from parabolic to elliptical brought about by a close approach of the comet to a major planet.

the circular velocity, the path of the body will be a parabola. At higher velocities still, the body will move in a hyperbolic orbit.

Now ellipses are all closed curves and a comet that moves in an elliptical orbit will make periodic returns towards the Sun provided there are no disturbing influences that will throw it into a different orbit. Several cases are known in which a comet travelling originally in a parabolic orbit has been captured by a major planet and forced into an elliptical orbit (Fig. 9) and, of course, the reverse can also occur.

One of the difficulties associated with the determination of the type of orbit being followed by a comet is that it is by no means easy to distinguish between the three types of curve when a comet is at perihelion, and unfortunately this is the time when they are best observed. In general, a comet is seen to its best advantage in that part of the heavens just above the rising or setting Sun. This is because it becomes sufficiently bright only when it approaches within a certain distance of the Sun, usually within about twice the radius of the Earth's orbit, and the effect of twilight or dawn makes observation somewhat difficult.

F

The Constitution of Comets

Comets are usually thought of as brilliant objects with long tails, and for centuries they have been regarded with a mixture of awe and superstition. Such a picture is only partly true. Comets actually possess only one feature in common: they all have a nebulous appearance, which is indicative of a mainly gaseous structure. A tail is not always present, for reasons we shall discuss later.

Normally, we may distinguish three different parts of a comet: the nucleus, which is a bright, often star-like object; the coma, which, as its name suggests, is a luminous cloud surrounding the nucleus; and the tail which, if present, stretches away from the head of the comet, sometimes for many millions of miles. Quite often it is difficult to make a distinction between the nucleus and the coma. On occasion the former is barely visible even in large instruments, and instances have been known when the nucleus has appeared to be multiple. In the same way, the coma may be little more than a luminous patch, whereas at other times it shows an extremely complex structure, very suggestive of a tail in the process of formation. Even the tail of a comet, when visible, can show remarkable variations, depending mainly upon the perspective as seen from the Earth. Several comets have been observed in which the structure of the tail has varied appreciably over a period of a few hours.

Quite clearly, from their appearance alone, the head of a comet does not consist of a single solid body. Further confirmation of this is obtained whenever a comet occults a star, since the latter may be seen quite easily through the head. From this we may conclude that a comet is little more than a vast nebulous mass, and estimates of density of comets show that, although their heads may be a million miles in diameter, their mean densities are much lower than that of the best vacuum obtainable on Earth. This is an extremely important fact, for it explains several of the spectroscopic features of comets.

The light from a comet comes from two distinct sources: first, reflected sunlight, and second the intrinsic luminosity of the comet itself, which forms the larger part of the light we receive. This raises an interesting point. Clearly, a comet does not possess any heat of its own and the glow we perceive must be produced by some other means. This mechanism is well known. Ultra-violet light from the Sun knocks out some of the electrons from the constituent atoms and subsequent rearrangement of the remaining electrons results in the emission of visible light. The spectroscope reveals the presence of several of these ionized atoms or molecules in different parts of a comet, particularly in the tail, where both ionized carbon monoxide and ionized nitrogen have been observed. The heads of comets, on the other hand, contain neutral molecules such as carbon, cyanogen and various hydrocarbons, together with certain radicals such as OH and NH. Hertzberg has also recently identified the presence of a triatomic molecule, CH_2, in the region close to the nucleus of certain comets.

All of these molecules are extremely reactive chemically and under ordinary conditions on Earth can only exist for extremely short periods. How then are they able to exist for much longer periods in comets? The answer lies in the point already made, that the density of the cometary substance is extremely low. Even within the nucleus, which is the densest part, a molecule can travel more than 500 miles without encountering another molecule, which means that virtually all of the molecules are moving independently of one another and can remain in this highly active state for quite long periods. Within the tail the density is even lower; here a typical molecule can travel a million miles without colliding with another.

The frequent appearance of the nucleus as a stellar point prompts the remark that it is here that any solid part of a comet exists. Although the precise nature of the nucleus still presents something of a mystery, observation with the most powerful telescopes indicates that a typical nucleus is extremely small in comparison with the rest of a comet, possibly not more than a few hundred yards in diameter and, since it is completely invisible whenever a comet passes in front of the Sun, it must consist of many small bodies held together by gravitation and the binding effect of the gases that are also present. The coma and the tail are completely gaseous. Confirmation of the solid nature of the nucleus is obtained from the fact that if a comet disintegrates – as several have – under the influence of the planets, debris is scattered throughout the orbit, such debris often appearing as a brilliant shower of meteors whenever the orbits of the Earth and the comet intersect.

The fact that the tails of comets always point away from the Sun as they swing around at perihelion has been known for several centuries. The first rational explanation of it was given in 1617 by Kepler, who suggested that some form of repulsive force emanating from the Sun drove the gaseous matter within the head outward, thus forming the tail. We now know that there are two different forces arising in the Sun which produce the tail of a comet. The first is ordinary radiation pressure which, although extremely small when applied to normal objects, becomes a dominant force upon particles in the vicinity of the Sun which are less than a thousandth of a millimetre in diameter. More recently, the effect of the solar wind – streams of electrified particles emitted by the Sun, especially by solar flares – upon the gases of comets has been shown to be equally effective in propelling the gases away from the head of a comet. Solar heating can also cause evaporation of cometary gases and, since all of these forces increase with proximity to the Sun, we have an explanation of why a comet does not show a tail until it approaches within a certain distance of perihelion.

We must also consider the loss of this tail-forming material from the heads of the periodic comets. Quite clearly, such a process cannot continue indefinitely. Since there is only a limited amount of this gas available for the formation of a tail, a time will come when all of it has been used up. Once this happens the head of the comet will begin to disintegrate during successive approaches to the Sun (or even at close approaches to the major planets), and it will then disappear completely, leaving only debris scattered along its orbit. The most prominent instance of this is Biela's comet, with the extremely short period of only 6·6 years, which in 1846 was found to have split into two components. At its return in 1852, both components were still travelling together, but since that time nothing has been seen of the comet itself, although a brilliant meteor shower occurs every six years or so, whenever the Earth intersects the orbit of the comet. Since several meteors are also observed from the same radiant every year, it is apparent that much of the detritus has now spread itself around the entire orbit.

Before leaving the question of comets, we must finally consider the case of those that travel in parabolic orbits. These comets make one appearance and are then never re-observed. Whether any comet really moves in a hyperbolic orbit is still open to question since those which, in the past, have appeared to do so from observations made at perihelion now seem to be moving in a parabola whenever we trace their motion outside the orbit of Pluto. Quite a large number of these non-periodic comets are known and the question naturally arises of their origin and their eventual destination once they leave the solar system. Many astronomers believe that a zone of comets exists far beyond the orbit of Pluto, possibly a light-year or so away, and that it is from this region that the parabolic comets are brought to perihelion, possibly by the perturbing influence of the nearby stars. At the present time, however, this question cannot be regarded as being definitely settled.

The Origin of the Solar System

Having described the different components that go to make up the solar system, we naturally wish to know how they came into being.

In the next chapter we shall be considering the origin and evolution of the stars, and it may seem strange at first sight to find that the problems of stellar origin and evolution is more readily susceptible of investigation and solution than that of our own planetary system. After all, the planets, their various satellites, the asteroids and the comets all approach relatively close to the Earth, whereas the stars lie at tremendous distances. There is, however, one fundamental difficulty associated with any theory of the formation of the solar system, namely that whereas there are stars by the million at almost every phase of their evolution available for observation, we have only one planetary system to examine in detail and we are at present trying to follow a trail that is more than 4,000 million years old. So many features of the planets are available for discussion that it is extremely difficult to determine which are either relevant or significant when it comes to postulating how they came into being.

Any theory we may put forward must explain the following characteristics of the solar system. The planets, with the exceptions of Neptune and Pluto, all show a mathematical relationship in the radii of their orbits (Bode's law) and apart from Mercury and Pluto they all have nearly circular orbits. The planes of their orbits all lie close to the mean plane of the solar system, the only exceptions being those of Pluto and certain of the asteroids and comets. All of the planets apart from Uranus and certain planetary satellites revolve and rotate in the same direction, and virtually all of the angular momentum of the solar system resides in the planets. To these conditions we may add a fifth, namely, that both the dimensions and the physical constitutions of the planets show a definite correlation with their distance from the Sun.

One of the first theories to be proposed was that of Kant and Laplace, a theory that was later modified by Roche and other astronomers during the nineteenth century. This supposed that the Sun condensed out of a vast nebulous cloud of gas and was originally a large red giant, very diffuse and much larger than at present. As it evolved, it contracted and rotated more rapidly, so that rings of gas were successively ejected which then condensed into the planets. Although this idea has now been abandoned in its original form, since it failed to explain the inclinations of the planetary orbits to the solar equatorial plane or the retrograde motions present within the solar system, it has recently been revived, notably by Hoyle.

An alternative mechanism was first propounded as long ago as the mid-eighteenth century by Buffon, on the assumption of a tangential collision of the Sun with another star. In essence, this idea – and particularly the later modification of it by Jeans, who

85

suggested a close approach by a second star rather than an actual collision – has many attractive features. For example, we can show that the tremendous tides that would be raised on the surfaces of each star when at minimum separation would give rise to jets of material ejected from the crests of the tidal bulges. As the stars moved farther apart, the bases of such jets would fall back into the parent bodies but the remainder would form a cigar-shaped mass of gas which, since it would have a fairly high rotational velocity, would break up into blobs of matter, and these on condensing would form the planets, which would continue to revolve around the parent star.

Certainly the hypothesis will account for certain features of the solar system. For example, it explains why the sizes of the planets are distributed as they are, with the giants near the centre of the spindle-shaped jet and the smaller ones at its extremities. It also provides us with the reason for the mean plane of the solar system not being the same as that of the Sun's equator, since the approaching star could have come in from any angle. Unfortunately, such a theory does not provide any explanation at all of the distribution of the planetary orbits in the form of Bode's law, an empirical mathematical relation giving the relative radii of the individual orbits.

Such an explanation has recently been provided by Weizsäcker, who has taken as his starting point the assumption that at some time during its early evolution the Sun possessed a gaseous, disc-like envelope. How this envelope came into being is a problem that we shall discuss later. Now this disc, which probably had a mass about one-tenth that of the Sun, would be unstable and the lighter gases, particularly hydrogen and helium, would dissipate quite rapidly into space. Present estimates suggest that the dissipation of these gases would take place in a period not longer than 60 million years. The remaining material would still be unstable and, as Weizsäcker has shown mathematically, would tend to develop into whirlpool-like cells, the centres of which would rotate around the Sun in circular orbits.

The importance of Weizsäcker's theory is that the dimensions of these cells will be controlled by the presence of their neighbours and, if we make the arbitrary assumption that each concentric ring contains five such cells, we find that the radii of the circles which are the boundaries of these rings are successive powers of 1·894, which describes the radii of the planets' orbits even better than Bode's law does. There are difficulties; for example, it is found that a larger number of cells must be postulated for Mercury, Venus and Pluto. However, this discrepancy may be due to the relatively greater instability of the gaseous disc at the extremities. One consequence of this hypothesis is that the planets will be formed by the interaction of two such cells at the point where their directions of rotation oppose each other; because of this, it can be shown that the planets must rotate in the same direction as they revolve around the Sun. The formation of the various satellites is regarded as being due to local eddies within the major vortices.

A development of the Kant and Laplace hypothesis has been suggested by Kuiper, who believes that the nebula surrounding the Sun and containing sufficient mass for the formation of all the solar bodies contracted and in doing so became unstable, breaking up into discrete clouds, which may be looked upon as protoplanets. We have already come across this idea of protoplanets in the discussion on the common origin of Mercury and Venus, and the Earth–Moon–Mars system.

Let us here examine the distribution of density throughout this gaseous nebula at the time when it began to break up into the protoplanets. We have already seen that much of the lighter gases such as hydrogen and helium will have been lost during the very early stages of planetary development, a fact that manifests itself in the present composition of the planets which, with the exception of Jupiter and Saturn, are grossly deficient in hydrogen when compared with the composition of the Sun. It may be argued, of course, that the gaseous disc had a composition very different from that of the Sun but, since both were formed from the same nebula, this seems highly improbable. Estimates of the amount of hydrogen that escaped from the developing solar system yield a figure of approximately seven times the combined mass of all the planets.

If we now consider the material that is left we encounter some very interesting features. One of the most recent theories of the origin of the gaseous nebula from which the planets were formed, put forward by Hoyle, suggests that the Sun was born, along with many other stars, in a large shower, much as stars are being born at the present time within the nebula in Orion (see Plate IX). As we shall see in the next chapter, the Sun is a member of what are known as Population I stars. The full implications of this do not concern us here. What is important in the present context is that these are stars which formed from cosmic material that had been previously enriched with heavy atoms produced in supernova explosions, and it seems quite likely that at least one supernova occurred within this large gas cloud, scattering these heavy atoms throughout the mass of the gas.

One further feature of the Sun is that it is revolving far more

87

slowly than it should. From dynamical considerations alone, a condensing mass of gas that is also contracting will rotate more and more rapidly during this contracting phase provided there is no other force acting upon it. If we assume reasonable dimensions and a rotational velocity typical of stars that are now observed to be in their early stages of condensation (such as those in the Orion nebula), then by now the Sun ought to be rotating at least fifty times as rapidly as it is. Obviously, some other force has acted upon it at some stage of its development, and we now have a fairly good idea what this force was. As the Sun condensed, a stage was reached when its angular velocity became so great that a disc of gas formed around the equatorial zone, the material in this disc being thrust further and further away from the Sun while at the same time more and more of the rotational momentum was transferred from the contracting Sun into the rotating disc. In the following chapter, we shall see that there is evidence from infra-red observations of the very young T Tauri stars that, in this case, such a disc is in the process of formation.

At this point we encounter a serious difficulty. At some stage the disc would become quite separate from the Sun and somehow the rotational momentum has to be transferred across the gap that is formed. One way in which this might be done has been put forward by Alvén, who has suggested that a magnetic field surrounding the Sun will not only allow rotational momentum to be carried across such a gap but will also have the effect of reducing the rate of rotation of the Sun while at the same time forcing the gaseous material further away.

Now although this process works perfectly well with gases, it will not do so with solids, and consequently we would expect the denser material such as iron and the various silicate rocks to remain in a zone quite close to the Sun, the lighter gases being forced away to greater distances. One great asset of the theory is that this is exactly the situation we now find. The inner, terrestrial planets are those containing a high proportion of rock and iron; the more distant giant planets are preponderantly composed of methane, ammonia, carbon dioxide and water.

On this view, the planets were formed, in the cold, from much smaller bodies; if we assume that the region where the last of the rock and iron gave way to the lighter gases lies somewhere beyond the orbit of Mars, then this accounts for the presence of the asteroid zone, for it would be here that the amount of available heavy solid particles was not sufficient for the formation of a body of planetary size. We might also add that, as Hoyle has pointed out, the densities of the satellites also fit in well with the above-outlined sequence of events, since we find that as we go further from the Sun the densities of the various systems of planetary satellites gradually become smaller. Two problems that still remain to be answered are the peculiar axial tilt of Uranus and the high density of Titan. As far as the first problem is concerned, Hoyle has postulated that Uranus was formed by the coagulation of two bodies of almost identical mass, but so far no explanation of the anomalous density of Titan has been put forward. Quite possibly Titan is an asteroid that was captured by Saturn during a close approach. Mention has already been made of the eccentric orbits of certain of the asteroids and this would certainly account for the anomalous density of this particular satellite. Here we are not necessarily implying that at the time of its capture Titan was in the form in which we now find it. It seems far more likely that, as with other large satellites, a core of heavy material was first captured, which then drew smaller bodies to itself by virtue of gravitational attraction. If Titan had initially formed within the asteroid zone, it would have been by far the largest of these bodies.

At present our observations are insufficient to provide a definitive explanation of the birth of the solar system and its evolution during the early stages. In view of the difficulties implicit in trying to piece together events that occurred about 4,000 million years ago, this is scarcely surprising. Whether information gathered from future space probes will allow us to accept one of the foregoing theories to the exclusion of the others, or whether a compromise will be necessary, remains to be seen. It may be that the vital clue will come from a different quarter, namely the study of planetary systems revolving about other stars. All that we can do at the moment is give a broad outline of the various ideas that agree reasonably well with the observed facts.

4. Stellar Evolution

The question of how the Sun has evolved to its present state and how it will continue to evolve in the future undoubtedly aroused the curiosity of astronomers long before they considered the way in which stars evolve, since it is only comparatively recently in the history of astronomy that the true nature of the stars, as suns in their own right, has been recognized. The Sun, being the nearest of the stars, is the only one to exhibit a measureable disc. Light, travelling at over 186,000 miles every second, takes slightly more than eight minutes to reach us from the Sun, but over four years to cover the distance from the nearest star system, Alpha Centauri. At first sight this would suggest that, if we wish to discover how the stars evolve from the moment they are born to the time they finally die, the Sun is the best candidate of all for us to examine. Certainly we can learn a great deal from a study of the Sun, but when it comes to predicting future events several million years from now, or the way in which the Sun behaved at a similar period in the past, we are forced to turn to the stars in order to obtain this information. A little reflection will show us why this is so.

Virtually all of the information we can gain from observation of the Sun is of what is actually happening now. If we had only the Sun to examine, we would not get very far in our study and any extrapolations we made would be hazardous in the extreme. When we come to examine the stars, however, we find representatives at all stages of evolution, from the youngest to the oldest, and from their characteristics we are able to tell, with a fair degree of accuracy, just what the Sun was like many millions of years ago and how it will behave millions of years in the future.

First, however, let us confine ourselves to a detailed examination of the Sun and of the various processes that are going on inside it, for much of what we learn here will help us to understand the reactions taking place within the stars. As stars go, the Sun is quite an ordinary body. Its diameter of 886,000 miles lies somewhere in the middle of the range of sizes found among the stars, as also does its mass. The surface temperature, too, of 5,750°K is not extreme by stellar standards. All of these features of the Sun are capable of direct measurement. But what is the interior of the Sun like, what kind of temperatures, pressures and nuclear reactions exist there beyond the range of direct measurement?

These are questions that obviously cannot be answered by mere observation. Instead we must rely upon calculations made by the astrophysicists, and we now know a great deal about conditions within the Sun and in particular about its composition and the various nuclear reactions that keep it shining at a constant rate for millions of years. There are several reasons why we would expect both the temperature and the pressure to rise as we progress towards the centre of the Sun. For example, we know that heat will always flow from a region of high temperature to one of lower temperature and, since heat is being continuously radiated from the surface of the Sun, it follows that the interior must be hotter than the exterior regions. In other words, a temperature gradient exists from the centre to the surface. Recent calculations indicate that at the centre the temperature must be of the order of 13,000,000°K and the density about fifty times that of water. These extremely high temperatures and densities are necessary for the central regions to withstand the pressure exerted by the overlying layers of gas. If by some means (which, as we shall see in the next chapter, can sometimes occur although it is unlikely in the case of the Sun in the near future) the pressure within the central regions were suddenly removed, there would be an immediate collapse which would release energy and thereby raise the central temperature. Such a rise in temperature would increase the pressure once more and a point would be reached where the collapse would cease as the pressures balanced once again. The result would be that for some time the Sun would pulsate around a mean size with a period of a few hours until, in the course of some centuries, a fresh stable balance would be struck, with the Sun hotter than it is at present, somewhat bluer in colour, and about half its present diameter. Later in the chapter we shall be discussing stars that are now in the process of undergoing these very regular pulsations, these being the variable stars known as the Cepheids and the RR Lyrae variables.

This balance of internal pressures is only one feature of an evolving star such as the Sun. Another, which is of equal importance, is that between the amount of energy that is radiated away from the Sun's surface and the energy production in the central regions. If the rate of energy dissipation into space were to fall below that being produced within the Sun, a slow expansion would occur with a resultant decrease in the internal temperature and a reduction in energy generation inside the Sun, until once more a balance would be attained. Indeed, the situation would be very similar to that just discussed, except for one important point. Such an expansion will be a slower, more sedate process, taking place over some millions of years.

Energy Flow within the Sun

Finally we must examine the means by which energy flows throughout the entire mass of the Sun. Deep within the central regions, energy flows by the process of radiation; a small fraction of this radiation is in the form of ultra-violet light, but most of it is of even shorter wavelengths (X-rays), which are extremely energetic. Now the energy associated with X-rays is so great that they are very efficient in removing electrons from atoms, knocking them completely from their orbits about the central nuclei.

The resulting atoms are said to be *ionized*. In other words, they have lost most, if not all, of their electrons (negative charges) and are therefore positively charged. Here, then, we have free electrons and stripped nuclei all moving around at very high velocities so that, in spite of the tremendous pressure, the material deep within the Sun can still be considered a gas. In the outer levels of the Sun, immediately below the photosphere, which is the bright disc that we see with the naked eye, energy is transported by an entirely different process, known as convection. This is the type of energy transfer we find in boiling water and we can indeed regard this region as actually boiling. This boiling gas can be seen by direct observation, since it forms large 'bubbles' or cells several hundreds of miles in diameter, which give the solar surface a peculiar granular appearance.

At the photosphere itself, energy transfer is once more by radiation. This is because convection requires matter in order to transport energy from one point to another. Only radiation can carry energy through empty space. To sum up then, there are three main regions within the Sun: (Fig. 10) an inner region where radiation by means of high-energy X-rays transports the energy outward, an intermediate convection zone, and finally the photosphere where radiation in the form of ultra-violet, visible and infra-red light carries the energy out into space.

The important consequence of this is that there is always a temperature gradient inside the Sun – and in any other star – with energy flowing from the central regions (where it is produced by well-understood nuclear reactions) towards the surface, and it is this more than anything else that governs the actual surface temperature, since the energy emitted by the photosphere must equal the flow from the energy-generating central zone. These three very important balancing mechanisms operating within the Sun and the stars prove

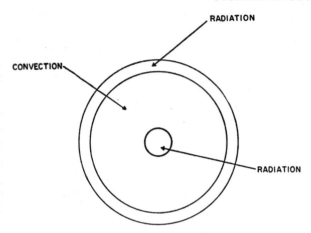

Fig. 10 The three regions of energy transfer within the Sun.

extremely useful in interpreting the various stages of evolution through which a star passes. It is to this that we shall now turn.

The Hertzsprung-Russell Diagram

In order to begin our discussion of stellar evolution, we must first look at two properties of a star that are readily observable. These are the brightness, which is directly observed, and the surface temperature, which is easily calculated from a study of the distribution of energy in the spectrum with respect to wavelength. Using these two characteristics, we may construct a diagram like that shown in Fig. 11. This was first done by Hertzsprung and independently by Russell, and the result is known as the Hertzsprung–Russell, diagram. As we need a standard luminosity, we take that of the Sun as unity.

The curve shown in Fig. 11 is known as the main sequence and represents the positions of those stars which have essentially the same chemical composition but different masses. We shall see later where other stars appear on the Hertzsprung–Russell diagram when they have different compositions from those lying on the main sequence. For the moment, however, we shall confine ourselves to an examination of the main-sequence stars, of which the Sun is a typical example.

The stars on the main sequence are those which have the same initial composition in which nuclear reactions going on in the centre have not modified the composition in that region to any great extent, and so long as this composition does not change radically, the star

93

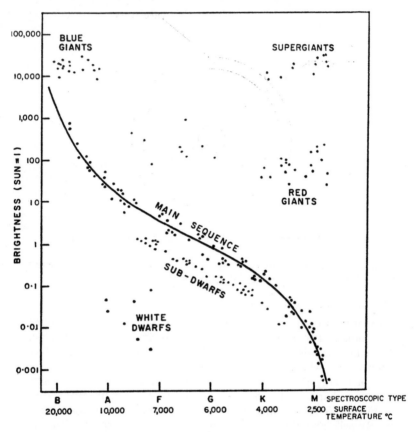

Fig. 11 The Hertzsprung-Russell diagram, illustrating the main sequence and the relative positions of various classes of star.

will remain on the main sequence, moving slowly up it as time goes on. During its stay there, certain nuclear reactions occur in its interior providing the star with energy.

Now so long as the star has a luminosity less than ten times that of the Sun, the dominant nuclear process is that known as the proton-proton chain reaction, which consists of three stages and requires only hydrogen and helium nuclei (the two lightest chemical elements) to be present:

$$^1H_1 + {}^1H_1 = {}^2H_1 + e^+ + \nu \tag{1}$$
$$^2H_1 + {}^1H_1 = {}^3He_2 + \gamma \tag{2}$$
$$^3He_2 + {}^3He_2 = {}^4He_2 + 2{}^1H_1 \tag{3}$$

94

In the initial stage, two protons combine to form a deuteron, the isotope of hydrogen that is found in heavy water. During this reaction one of the protons changes into a neutron, emitting at the same time a positron (a particle having the same mass as an electron but with a positive charge), this particular kind of reaction being known as β-decay. The second stage consists of the addition of a further proton to the deuteron with the formation of an isotope of helium, which has a nucleus containing two protons and a neutron, radiation being given out at this stage in the form of γ-rays. Finally, two of these helium nuclei combine to yield an atom of normal helium, $^4He_2$, with two protons being ejected. Inside a star such as the Sun, for example, all these reactions are going on simultaneously, the overall result being the conversion of hydrogen into helium. It will be noticed, too, that energy is produced during all of these stages, in the form of the positron, the γ-radiation emitted during the second reaction, and finally the energy of motion of the two ejected protons.

So far, we have made no mention of the particle shown as v that is emitted along with the positron during the initial β-decay process. Later in the chapter we shall see how this particle, which is known as the neutrino, plays an important role in the complex series of reactions leading to a supernova explosion, and it is perhaps worth while here to discuss some of the properties of the neutrino in detail.

The Neutrino

A large number of nuclear reactions are known in which a proton changes into a neutron, and careful measurements have shown that the positrons emitted do not move with the same velocity. In order to explain this, Pauli postulated that a second particle is also emitted which carries away with it a portion of the energy released by the decay of the nucleus. Although the total amount of energy released is always the same, it is divided unequally between the positron and this second particle, the neutrino. Unfortunately, the neutrino possesses neither charge nor mass and is consequently extremely difficult to detect.

Theoretically, there are several nuclear reactions in which the neutrino can take part, the most useful being that in which a certain isotope of chlorine is converted into an atom of argon, a rare gas which can be readily separated from the reaction mixture and accurately measured. Although this reaction has not yet been successful in detecting neutrinos from the Sun, the negative results may be an indication that the central temperature predicted by other methods is a little too high. Since it has neither mass nor charge the neutrino passes unhindered out of the stellar interior, removing some energy with it, energy that is completely lost into space. In a small number of stars, this loss of energy by neutrino emission exceeds that by radiation from the surface, resulting in a complete and catastrophic collapse.

Where, in the star, do these reactions take place? When a star is born, it is composed almost entirely of hydrogen, with only very small quantities of helium, certain metals, carbon, oxygen, nitrogen and other gases. We would be quite wrong to suppose that the reactions just outlined begin at any arbitrary point within the mass of the star. Both temperature and pressure have to be of the right magnitude for the processes to begin and it is in the centre of the star that conditions are right. What happens then is that helium is gradually formed at the centre by the reactions just described, or others very similar which will be discussed later. Evidently this helium may remain where it is or mix with the remaining hydrogen but, before we discuss this very important point, we must first consider an alternative series of nuclear reactions that go on in those stars which are more than ten times as luminous as the Sun. It will be recalled that earlier we said the proton-proton chain reaction was dominant in those stars up to a tenfold solar luminosity. In the case of the more luminous stars, the carbon-nitrogen cycle, first discovered by Bethe and independently by von Weizsäcker, is of much greater importance, since here the temperatures are sufficiently high to allow this more complex series of reactions to take place.

The Carbon-Nitrogen Cycle

This series of reactions accounts very well for the energy output of the very luminous stars, ranging in spectral type from O to F, and it is significant that we find the abundance of carbon and nitrogen in these stars to remain virtually constant throughout the entire spectroscopic series between these limits. Although, as we have said, this cycle is more complex than the proton-proton chain reaction, the net result is still the same, namely the conversion of hydrogen into helium. The full sequence is as follows:

$$^{12}C_6 + {}^1H_1 = {}^{13}N_7 + \gamma, \tag{4}$$
$$^{13}N_7 = {}^{13}C_6 + e^+ + \nu, \tag{5}$$
$$^{13}C_6 + {}^1H_1 = {}^{14}N_7 + \gamma, \tag{6}$$
$$^{14}N_7 + {}^1H_1 = {}^{15}O_8 + \gamma, \tag{7}$$
$$^{15}O_8 = {}^{15}N_7 + e^+ \nu, \tag{8}$$
$$^{15}N_7 + {}^1H_1 = {}^{12}C_6 + {}^4He_2. \tag{9}$$

Quite clearly, both the carbon and the nitrogen may be used over and over again in the reactions; they are termed nuclear catalysts, since they are virtually unchanged. Stages (5) and (8) are both β-decay processes and, as before, neutrinos are emitted, removing energy from the star into space. In all of the other reactions, the energy produced is absorbed by the star.

Mixed or Unmixed Stars

At this point we must consider the two alternatives mentioned earlier. Does the helium, as it is formed, mix with the hydrogen so that as time goes on the helium content increases but the composition of the star remains essentially iniform throughout? Or does the helium remain where it is, in the centre, forming a slowly increasing core with little, if any, mixing? This turns out to be a very important distinction as far as the evolution of the star is concerned. Calculation has shown that, if the two gases become intimately mixed, the star will evolve along the paths shown in Fig. 12, moving slightly to the left of the main sequence but not departing very far from it.

This was the situation as envisaged by astronomers during their initial studies of the evolution of stars. More recent investigations, carried out in particular by Mestel, have thrown considerable doubt on this theory and it is now believed much more likely that scarcely any mixing at all occurs. In this case, the course of evolution is far more dramatic, as may be seen from an examination of Fig. 13. Here, the later stages of a stars' life are spent well away from the main sequence; the star moves first toward the right and then, after reaching point C, makes an abrupt turn-around, evolving back towards the main sequence, crossing it at a point higher than when it left it, and finally entering the region of the white dwarfs.

It is important here to realize that the rate at which a star moves along the different portions of the curve shown in Fig. 13 is far from uniform. For example, evolution from C to D takes only about one-hundredth of the time required to go from A to B. At the moment the Sun is moving from A to B; it has taken some 5,000 million years to reach its present position and will require approximately the same length of time to reach point B. Once it begins to evolve towards C, however, its evolution will become accelerated and events will occur far more rapidly.

Another factor that determines the rate of evolution in a star is its mass. The more massive a star, the more rapidly it will evolve; this is shown very clearly in the case of the globular clusters which contain many thousands of stars of widely differing masses. Present-

G

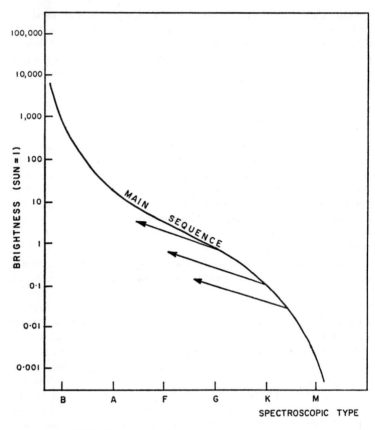

Fig. 12 Evolution of mixed stars from the main sequence.

day astronomical theory indicates that all of the stars in a globular cluster were born at about the same time. Fig. 14 shows the distribution of the stars in one particular cluster on a Hertzsprung–Russell diagram. The similarity between Figs. 13 and 14 is readily seen. Many stars are still on the main sequence. These are stars with masses very like or less than that of the Sun. Those that have a slightly greater mass, and possibly are just a little older, have already begun to evolve to the right of the main sequence. Others that are even more massive have now passed the sharp turn-about at *C* and are evolving back towards the left of the diagram. This direct observational evidence is a very strong point in favour of the theory that no mixing occurs within the stars.

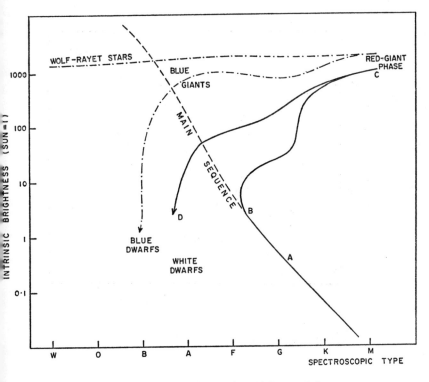

Fig. 13 Evolutionary track of a star in which no mixing occurs.

Nuclear Reactions and Stellar Evolution

We have now reached the stage where it is possible to determine the effect that the various nuclear reactions taking place within a star have upon the different phases of its evolution. In particular, we shall see why it is that evolution takes several thousand million years to reach the point C, whereas the later stages (which are also closely related to stellar variability), succeed each other quite rapidly. During the initial stages, a typical main-sequence star produces helium in its centre, where both temperature and pressure are greatest, either by the proton-proton chain reaction or by the carbon-nitrogen cycle, depending upon its luminosity. After a time, however, all of the hydrogen in this central region will have been converted

99

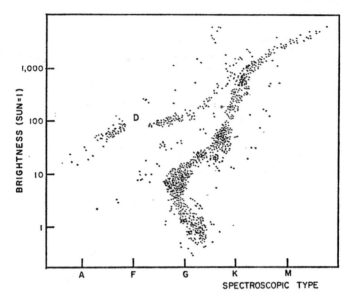

Fig. 14 The Hertzsprung-Russell diagram of stars in a typical globular cluster.

into helium and a core of this material will have been formed. Thereafter, a relatively thin energy-generating skin is formed around the core while, at the same time, the core begins to shrink in order to maintain the production of energy.

Although we might expect the whole star to contract during the shrinkage of the core, this is not the case, and it actually swells as it moves away to the right of the main sequence. It now seems quite well established that the formation of a shrinking helium core within the star brings about its departure from the main sequence. During this phase, energy transfer within the star is just as we have already described: first, radiation in the centre, followed by convection in an outer layer as the dominant process, and finally radiation again at the photosphere. Within the central core of helium, however, the energy flow ceases after a time; since this is an estremely important point, we shall discuss it here in some detail.

100

Degeneracy Pressure

The point has already been made that the central temperature of the Sun is very high, probably several million degrees, and the pressure here due to the weight of the overlying layers is also extremely high. The material in the centre, however, still behaves like a gas in spite of these extreme conditions, for both electrons and atomic nuclei have perfect freedom of movement and are not constrained in any way as they are in a liquid or solid. Once the central core of helium begins to shrink though, something very peculiar happens. The density rises far in excess of that now present in the centre of the Sun, rising so high, in fact, that most of the atomic nuclei are completely stripped of their electrons. A normal atom would occupy far too great a volume of space to exist in such an environment. Such material is therefore composed of a heterogeneous mixture of electrons and atomic nuclei, the energy which the eletrons originally possessed to maintain them in their orbits now being translated into random motion. This strange condition was first investigated theoretically by Fermi who was able to show that such material (which is known as degenerate matter), exerts a powerful pressure of its own (degeneracy pressure) owing entirely to the energetic random motions of the particles.

The rise in this degeneracy pressure inside the star can be quite rapid and, unless a very critical condition is maintained, will lead eventually to almost total disruption of the star in the form of a supernova explosion. This condition is that the mass of the helium core must not exceed 1·44 times the mass of the Sun, which is known as Chandrasekhar's limit. In the following chapter we shall see what happens to stars in which this limit is exceeded.

In the case of the stars we are at present considering, where the mass of the helium core lies below Chandrasekhar's limit, the degeneracy pressure will eventually become sufficiently high to prevent the core from contracting any further and it will then attain a temperature that is primarily determined by the mass of the star and lies within the range from 15,000,000 to 20,000,000°K. By the time this has become established, the star will have reached the curve on the evolution diagram at *B*.

The star now commences to expand into a red giant so that, although its surface temperature falls somewhat, its total luminosity increases owing to the much greater surface area. Meanwhile, in the central regions, the temperature of the growing helium core rises slowly, accompanied by a large increase in the depth of the convection zone, which may eventually almost reach down to the expanding energy-generating skin around the core.

101

Shortly before the point C is reached on the curve, the central temperature has risen sufficiently for a new sequence of nuclear reactions to begin, reactions that result in the burning of the helium itself. Such high temperatures are necessary for, as physicists have shown, the helium nucleus is one of the most stable nuclei known. The net result is the formation of oxygen and neon:

$$^4He_2 + 2\,^4He_2 = \,^{12}C_6 + \gamma, \tag{10}$$
$$^{12}C_6 + \,^4He_2 = \,^{16}O_8 + \gamma, \tag{11}$$
$$^{16}O_8 + \,^4He_2 = \,^{20}Ne_{10} + \gamma. \tag{12}$$

By now, the central temperature is probably about $100,000,000°K$ and, since a large proportion of the material in the core is degenerate, the reaction gets out of hand and an explosive condition is set up. Under normal conditions, the helium burning and this high temperature would not, of themselves, lead to an explosion. It is the fact that the material is degenerate which leads to the energy generation's getting out of control. Now why should the fact that the material is degenerate have such a dramatic effect upon the nuclear reactions taking place? The answer lies in one very peculiar property of degenerate matter. Unlike an ordinary gas, it does not have the inherent ability to balance its temperature with the outflow of energy. In a normal gas, if the outflow of energy is less than that being produced by the nuclear reactions, the temperature falls due to adiabatic expansion. In the case of degenerate matter, the opposite happens. The temperature rises and more energy is produced, but owing to its peculiar equation of state it cannot release energy by adiabatic expansion and the temperature rises very quickly until degeneracy is removed.

The central core of the star explodes with considerable violence but since the core lies below the outer levels of the star, the explosive force is contained. The effect, known as the 'popping' of the core, brings about the abrupt turn-around at C and starts the star evolving back in the direction of the main sequence.

Contraction now takes place and the surface temperature rises correspondingly as more hydrogen is transmuted into helium and other nuclei act as fuels. A point is reached, however, when all of the hydrogen, helium and other nuclei have been used up; then a slow cooling begins, the surface temperature and luminosity decreasing in step with the overall contraction. During these late stages of evolution, degeneracy pressure in the interior of the star maintains the pressure balance (the mass is now well below Chandrasekhar's limit) and allows this extremely slow cooling to take place, eventually placing the star in the white dwarf region. These stars are one of the most common type found beyond the end of the evolutionary curve shown in Fig. 13.

Variable Stars

In our tracing of the evolution of a star, so far we have tacitly assumed that any changes that occur (with the exception of the 'popping' of the core following the onset of helium burning) are quite leisurely affairs, taking millions of years. For almost four centuries, however, certain stars have been known in which quite dramatic and rapid changes of brightness occur – the variable stars – and it is appropriate here to consider where these fit into the evolutionary scheme shown in Fig. 13.

The first such star to be discovered and intensively investigated was Mira (*o* Ceti), found to be variable by Fabricius in 1596 and shown by Holwarda in 1638 to vary periodically between second and tenth magnitude with a mean period of 331 days. The typical light variations of this star are shown in Fig. 15, where it will be readily seen that, although there is this basic periodicity, it is far from regular and, in addition, the individual cycles are quite different from one another. Indeed, it is quite impossible to predict the shape, amplitude or duration of any cycle from those which have preceded it.

Many thousands of variable stars are known at the present time, and an extensive study of their light variations and, to a somewhat lesser extent, their spectroscopic changes, has enabled astronomers to classify them into several different types. How these various classes of variable stars fit into the overall evolutionary scheme has still to be fully answered. Certain types fit in very well, while others are more difficult to assign to a particular region of the evolutionary curve. We shall begin this discussion with a description of those variables that appear to be among the youngest of all the stars.

The T Orionis and T Tauri Variables

Within the Galaxy we find several large gaseous nebulae, of which the great nebula in the constellation Orion is by far the finest example, being readily visible to the naked eye on a clear, moonless night (see Plate IX). Very little was known of their true nature before the pioneering spectroscopic work of Sir William Huggins who was able to show that the spectra of such nebulae contain essentially bright emission lines such as we would expect to obtain from vast masses of glowing gas. It is quite clear that the gas itself must be self-luminous although a high proportion of the emitted light comes from the radiation of very hot stars embedded deeply within it. Several of the lines in the spectra are of ionized atoms, the predomi-

103

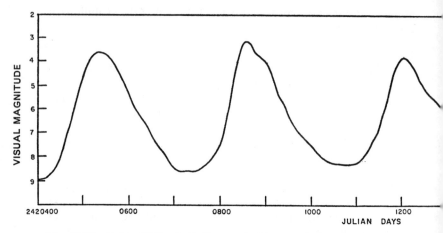

Fig. 15 The light of Mira (*o* Ceti), a typical long-period variable.

nant gases being hydrogen, helium, oxygen and nitrogen, providing further proof of the high surface temperatures of the embedded stars. These are predominantly of spectral classes O, B and A, with surface temperatures in excess of 12,000°K.

A very high proportion of these stars are variables, the prototype being the star T Orionis, normally of about magnitude 9·5 but fading irregularly and often abruptly by as much as three magnitudes, as shown in Fig. 16. There is now quite a lot of evidence that these variables are hot, giant stars forming out of these large gas clouds, stars with the high surface temperatures just mentioned.

Now one effect of these high temperatures is that much of the radiation will be in the ultra-violet and this results in a very steep temperature gradient between the star and the surrounding gas. Consequently, we would expect high pressures to be operative in the region around these condensing stars, the hot regions of gas forcing the cooler areas away, creating violent distrubances around the stars. By this means we are able to see quite a distance into the gas clouds where, although a fair part of the surrounding gas has been forced away, enough still remains for the process of star formation to continue.

It is generally agreed that the rapid and irregular light fluctuations which are so characteristic of these T Orionis variables are due to a combination of at least two processes going on simultaneously. First, there are inherent unstable fluctuations closely associated with

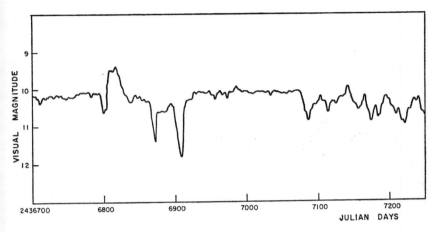

Fig. 16 The light curve of T Orionis.

the initial stages of star formation, and second, there are accompanying changes in the opacity of the surrounding gas and dust as they are forced away from the vicinity of these blue giant stars.

In addition to the stars themselves, certain quite small and discrete regions have been discovered in and close to the Orion nebula which, although invisible in optical telescopes, do show up fairly strongly in infra-red telescopes, and there appears to be little doubt that these are actually protostars just on the point of condensing out of the gas under the influence of gravitation, but not as yet sufficiently hot to radiate in the visible region of the spectrum. If this hypothesis is correct, then we would expect such protostars to become visible in a short time, possibly only twenty years or so.

The T Tauri stars are also young objects just forming out of dust clouds, but in this case the nebulosities with which they are associated are comparatively small and the stars themselves are not blue giants, but red dwarfs of low intrinsic luminosity. The faint nebulae from which these variables are condensing also vary in brightness showing that, in part, their luminosity is dependent upon the radiation they receive from the embedded stars. However, there is quite a lot of observational evidence which suggests that the light fluctuations of the nebulosities are greater than those of the stars themselves, indicating that the small gas clouds possess some self-luminosity. Unlike the light curves of the T Orionis stars, those of the T Tauri variables usually consist of fairly slow, sinusoidal waves with quite

105

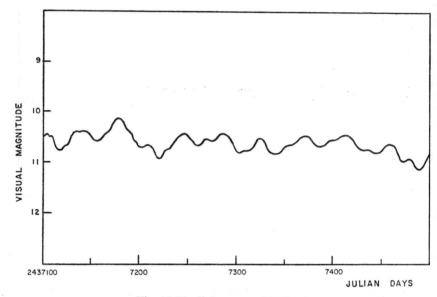

Fig. 17 The light curve of T Tauri.

small amplitudes, any pronounced variations occurring around the times when the star is bright (Fig. 17).

The spectra of these stars show broad, diffuse absorption lines which are suggestive of a high axial rotation, and no other stars close to the main sequence having such high rotational velocities are known. Herbig has suggested that these variables are still in the process of contracting under the influence of strong gravitational forces which may account in part for the rapid rate of revolution and, as we shall see in a later chapter, infra-red investigations carried out on certain of these stars indicate that here we may also have planetary systems forming. It is important also to bear in mind that the conservation of angular momentum may not hold in the case of such very young objects which are just condensing out of the pre-stellar dust clouds.

Now where do these variables appear on the Hertzsprung–Russell diagram? The bright T Orionis variables are to be found to the right of the main sequence, high up among the blue giants, and appear to be evolving rapidly towards the main sequence. The T Tauri stars, on the other hand, although again to the right of the main sequence and evolving towards it, lie to the bottom right of the Hertzsprung–Russell diagram. As they evolve they will become hotter, and smaller

106

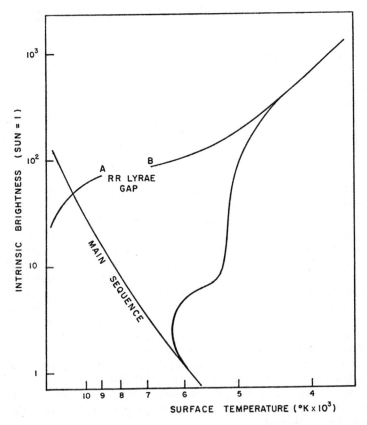

Fig. 18 Change of period across the RR Lyrae gap. These range from about eight hours at *A* to a little under a day at *B*.

by the time they eventually reach the main sequence. RY Tauri, for example, which is now a G-type dwarf, will reach the main sequence in about 10 million years, eventually becoming an A7-type star. In the process, conservation of angular moomentum will result in a doubling of its axial velocity to about 100 kilometres per second, a value, incidentally, which is common to similar A7 stars that are now on the main sequence. Those T Tauri variables which have later spectral types between K and M will evolve into G0 stars by the time they reach the main sequence and in their case the rotational velocity will decrease until they form stars that are very similar to the Sun. It is interesting, therefore, to speculate that the Sun may

107

once have been a T Tauri variable with an M-type spectrum, although this is extremely unlikely.

The RR Lyrae Variables

A close look at Fig. 14 shows that there is a gap in the evolutionary curve at the point D. This is known as the RR Lyrae gap since it is here that we find a class of variable star whose light variations are not only very characteristic, enabling these stars to be readily identified in large numbers, but also of great importance, for the RR Lyrae variables or, as they are sometimes called, the cluster variables, make it possible for us to measure large distances extremely accurately.

The periods of these stars lie between about eight hours and a little over a day, depending upon their position within the gap. As a star evolves towards the main sequence along the curve, the period decreases, as shown in Fig. 18. At present, we cannot say with certainty whether the oscillations induced in these stars are brought about by some disturbance of the pressure balance, although this seems likely. What is evident is that the onset of the oscillations is quite sudden, the star commencing to pulsate as it enters the instability gap from the right of the Hertzsprung–Russell diagram with a period of about a day, the oscillations being gradually damped as evolution progresses until, once the period has diminished to about eight hours, they die out altogether.

The time spent by the star within this gap is of the order of a few million years and accordingly, for any particular star, the pulsations as we observe them now are extremely regular, any change in the period being exceedingly small over the period during which they have been observed.

The light curves of the RR Lyrae stars show certain characteristics which enabled Bailey to classify them into three main groups (Fig. 19), depending upon their periods, the steepness of the rise to maximum brightness and the general shape of the light curve.

One further feature of the RR Lyrae variables is that there appears to be two distinct groups of these stars, with a sharp break in the periods at about ten hours (Fig. 20). So far the reason for this is not completely known, although it may be connected with what is known as the Blazko phenomenon associated with this particular type of variable. About 1906, Blazko made a detailed study of two of the RR Lyrae stars, RW Draconis and XZ Cygni, from which he found that both the amplitude and the shape of the light curves vary in periods of 41·72 and 57·24 days respectively. Since that time several other RR Lyrae variables have been found to exhibit the Blazko

108

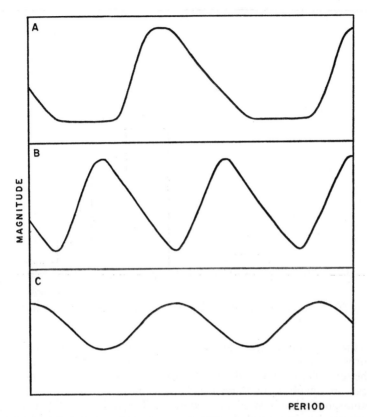

Fig. 19 The three classes of RR Lyrae variables according to Bailey's classification.

phenomenon. At first it was thought to be due to the interference of two or more oscillations going on within the mass of the star, but more recent investigations have revealed that the real explanation is far more complex than this.

We can, however, learn a lot from the spectroscopic changes that accompany the light variations, since these provide much basic information concerning the fundamental mechanism of the pulsations. From the Doppler shift of the lines in the spectra we obtain radial-velocity curves which show how the surface of the star is moving with respect to us, that is, whether it is approaching or receding as the light varies. Now the radial-velocity curves prove the existence of the pulsations in these stars, although the changes

109

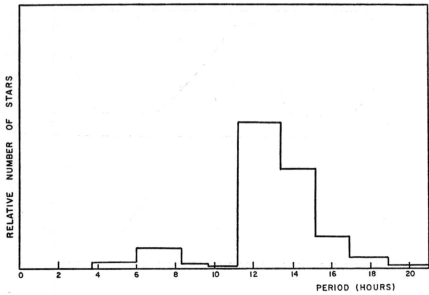

Fig. 20 Diagram illustrating the break at about ten hours in the periods of the RR Lyrae stars.

in the diameter necessary to account for the observed light variations turns out to be quite small, amounting to only about 10 per cent of the total diameter. It is doubtful whether the internal regions of the star are involved in these oscillations; probably only the outer atmospheric levels swell and contract. The radial-velocity curves indicate that, in general, the maximum brightness corresponds almost exactly to the maximum velocity of approach of the stellar surface.

One further feature of the RR Lyrae variables is worthy of note. A study of their spectra has shown that, if we calculate the spectral class from the intensity of the hydrogen lines that appear in absorption at minimum brightness, this is often later than that calculated from the strength of the calcium lines. This peculiarity has been examined by Zessewitch, who has shown from a study of more than 200 of these stars that those RR Lyrae variables which have a variable period (such as RW Draconis and XZ Cygni already mentioned) are deficient in their metal content as compared with those having a strictly constant period. Whatever the reason for this, it would seem that the overall stability of the pulsations in these stars is dependent in some as yet unknown way up on the relative abundance of metal atoms in the atmospheric layers.

110

A Measurement of Interstellar Distances

One extremely important characteristic of the RR Lyrae variables is the fact that they all have the same intrinsic brightness, their luminosity being about a hundred times brighter than the Sun. This means that they can be seen for great distances. Several are known, for example, which must lie on the very periphery of the Galaxy and others have been discovered in great numbers in the globular clusters and in the Magellanic Clouds, which are two satellite galaxies of our own, visible only from the Southern Hemisphere and some 180,000 light-years distant. More important, it means that we can utilize these stars to measure very great distances.

It is convenient at this point to say a few words about this method of distance measurement. The stars we see, either visually or on a photographic plate, are of widely varying brightness – or magnitudes – but unfortunately we cannot simply use the magnitudes to tell us how far away any particular star lies. If all of the stars were identical in their intrinsic brightness, then the fainter a star appears to us, the further away it would be, since the amount of light reaching us is inversely proportional to the square of the distance. Owing to the very wide range of luminosities found among the stars, however, this not possible. A dim star may be a red dwarf comparatively close to us or a blue supergiant several thousand light-years distant. Conversely, a first-magnitude star might be a nearby red giant, such as Betelgeuse, or a much more distant blue giant, such as Rigel.

The application of the RR Lyrae stars as distance indicators rests upon two facts: first, they are all of about the same intrinsic brightness and second, their characteristic light variations render then readily recognizable. The early investigations of these variables indicated that they have an absolute magnitude (the apparent magnitude at a standard distance of 10 parsecs or 32·6 light-years) of 0·0, and the distances of many of the globular clusters were estimated on this basis. More recent work, carried out by Chalonge and Miss Fringant, has resulted in a revision of this early value, the accepted figure of +0·8 being somewhat fainter than was originally thought. This has meant that the original distances of the globular clusters have been reduced somewhat, confirming other evidence which had accumulated concerning the size of the Galaxy.

In addition to providing us with an estimate of the size of the Galaxy and also of the distances to the Magellanic Clouds, Baade has used the RR Lyrae variables to measure the distance of the Sun from the centre of the Galaxy. The Sun and its attendant planets are

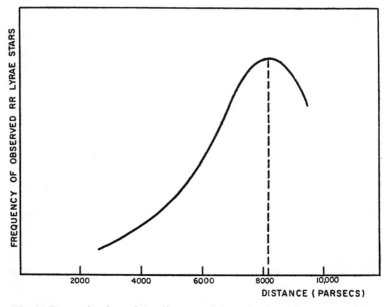

Fig. 21 Determination of the distance of the galactic centre using the RR Lyrae variables.

situated in an arm of the Galaxy some distance from the centre, the Galaxy itself having a lenticular structure with quite a pronounced bulge towards the centre. By measuring the distances of the RR Lyrae variables which lie in the direction of the galactic centre and comparing their frequency with distance we obtain a frequency-distribution curve of the type shown in Fig. 21, from which we see that the number of these stars reaches a maximum at a distance of just over 8,000 parsecs, this being the distance of the Sun from the centre of the Galaxy. All in all, the RR Lyrae variables have proved to be an extremely useful and powerful tool for the astrophysicists in their survey of the Galaxy and the surrounding halo of globular clusters.

The Cepheid Variables

The light curves of the Cepheids, which are named after δ Cephei, the brightest member of this class, are so like those of the RR Lyrae variables that for a long time they were classed together causing, as we shall see in a later chapter, a great deal of confusion in the distance measurements. From Fig. 22 it will be seen that the Cepheids

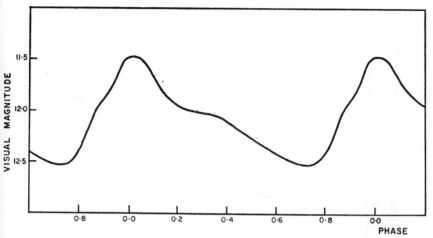

Fig. 22 Light curve of the Cepheid variable EV Aquilae, with a period of 39.5989 days.

are characterized by a rapid rise in light to maximum, followed by a slower decline. The period from one maximum to the next is generally very constant. Quite often there is a pronounced hump on either the ascending on the descending branch of the light curve, and Gaposhkin has demonstrated that such irregularities show a tendency to become more noticeable among those stars having long periods. The periods of these stars are all longer than those of the RR Lyrae variables and cover a far wider range, from about 1·5 days to more than 50 days. The record is probably held by DN Arae with a period of 82 days, although since this particular star is very faint there still remains some doubt as to whether it is a true Cepheid. The Blazko phenomenon is rarely, if ever, found among the classical Cepheids.

Unlike the RR Lyrae stars, the Cepheids are not even approximately of the same intrinsic brightness and were it not for a remarkable feature of these stars it would be extremely difficult to use them for the determination of distances. Early this century, Miss Leavitt made a very special study of the Cepheids in the Magellanic Clouds and discovered that the longer the period, the brighter was the star when at maximum (Fig. 23).

Had this discovery been made among the Cepheids in our own Galaxy, it would have meant very little since here the unknown distance factor has to be taken into consideration, but since the overall dimensions of the Magellanic Clouds are small compared with

H

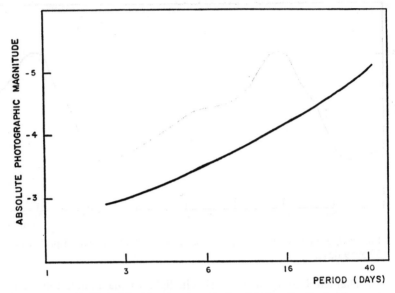

Fig. 23 The period-luminosity relationship for the Cepheids.

their distance from us, we may assume that all of the Cepheids in them are at essentially the same distance, and as a result their periods are also proportional to their intrinsic brightness (their absolute magnitudes). This relation between period and absolute magnitude is known as the period-luminosity law and consequently, if we know the period of any given Cepheid, which is readily obtained from its light curve, we may calculate its distance by means of the following simple equation:

$$M - m = 5 - 5 \log D \tag{13}$$

where M is the absolute magnitude, m is the apparent magnitude and D is the distance in parsecs.

In a later chapter, we shall see that certain difficulties arise when we attempt to use the Cepheids as distance indicators in practice.

Evolution of the Cepheids

Earlier, we saw how the RR Lyrae variables are found within an instability gap in the evolution curve, these all being stars that originally lay on the main sequence at a point only slightly higher than that now occupied by the Sun. Since the Cepheids are much brighter even than the RR Lyrae stars, we would expect them to have

114

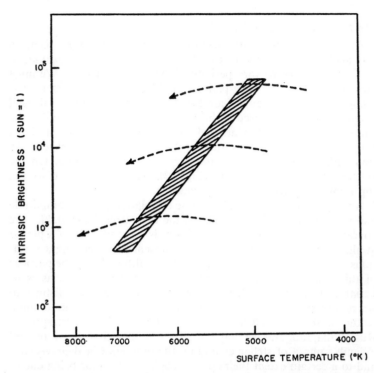

Fig. 24 The Cepheid zone. Stars at the lower end were initially as bright as the Sun when on the main sequence; those at the upper end were about 300 times brighter.

been initially main-sequence stars much more luminous than the Sun, originally between 10 and 350 times as bright.

The evolutionary track of these stars is very similar to that of the RR Lyrae variables but, since they are brighter and presumably more massive, they will evolve more quickly and also higher up the Hertzsprung–Russell diagram to the right of the main sequence. The Cepheid instability region is found to be a vertical zone, rather than a short horizontal gap, as shown in Fig. 24.

The interesting feature of the Cepheid instability zone is that both the period and the absolute magnitude increase from the lower end to the top, the period varying from about 1·5 days at A to around 60 days at B. As the period may be determined quite accurately from observation, we can place any particular Cepheid along the instability zone and read off its intrinsic brightness at once.

115

The Long-Period Variables

At the beginning of this section on the variable stars, mention was made of σ Ceti, the first such star to be extensively examined, which was shown to vary periodically in about 331 days. We must now see where these long-period variables and the closely related semi-regular and irregular variables lie on the Hertzsprung–Russell diagram, and also discuss their probable evolution. Unlike the variable stars already mentioned, these variables are all red giants with low surface temperatures (which vary almost exactly in step with the light changes) and very low mean densities. On the Hertzsprung–Russell diagram (Fig. 13), we find them far to the top right, away from the main sequence.

Among these stars we find no close correlation between amplitude and mean period; indeed, as far as the truly irregular variables are concerned, as their name suggests, there is no semblance of a period at all, every trace of periodicity being absent from their light variations. We must therefore bear in mind that although, for the purpose of this discussion, it is convenient to treat these three classes of variable star together, they are nevertheless sufficiently distinct to warrant their being divided into individual types. Their one common feature is that they all lie in the same region of the Hertzsprung–Russell diagram and we may tentatively assume that they represent a similar phase of stellar evolution, the difference in their visual – and to a certain extent their spectroscopic – behaviour being one of degree rather than kind.

As we have seen, their surface temperatures are continuously changing, such variations being reflected in their light curves. Like the RR Lyrae and Cepheid variables, they are believed to be pulsating stars, the surface temperature decreasing as the star expands and rising again as it contracts. For one or two of these stars – for example, Betelgeuse and Antares, both of which are irregular variables – we have direct evidence of this. Both of these stars are red supergiants and are sufficiently close to us, 586 and 424 light-years respectively, for their diameters to be measured by means of an instrument known as an interferometer. Such measurements have proved conclusively that their dimensions do indeed change that of Betelgeuse, for instance, varying between 300 and 400 solar diameters.

Let us see now if we can learn anything of the reasons for this variation in brightness from the spectra of these stars. The long-period variables, in particular, have provided astronomers with a tremendous amount of spectroscopic information, and we find that

116

all of these stars (and the semi-regular and irregular variables) have spectra of classes M, N, R or S, indicative of low surface tempera- tures, often with bright emission lines of hydrogen that appear around the times of maximum brightness. The major feature of the spectra of these stars is the presence of strong absorption bands due to molecules which are stable at the relatively low temperatures prevail- ing in these stellar atmospheres, particularly those of titanium and zirconium oxides and certain carbon compounds. Since these strong absorption bands appear in the visible region of the spectrum, much of the dimming of the light is due to the veiling effect of these bands. This is clearly shown when we compare the output of visible light with bolometric estimates that take into account the total radiation from a star, including both the ultra-violet and infra-red which are absorbed by our atmosphere and do not penetrate down to the surface. Visually, a typical long-period variable brightens about 100,000 times from minimum to maximum, whereas bolo- metrically the increase in the total radiation is only doubled.

At present, our knowledge of the basic cause of the pulsations in these variables is extremely meagre. As several thousand are known today, it appears reasonable to suppose that this variabliity of the long-period stars is an evolutionary phase of a star prior to the Cepheid stage. We shall have more to say about this later when we consider the picture of the variable stars as a whole. In the meantime, let us examine two effects that certainly contribute to the light variation, namely, the pulsation of the outer levels of the star and shock-wave phenomena within its atmosphere.

The pulsations occurring in the long-period variables may be readily distinguished from those of the Cepheids and RR Lyrae variables since in the latter the periodicity is maintained with an almost clockwork precision throughout successive cycles, whereas no two cycles of the long-period variables are identical and the interval between two maxima varies quite widely about the mean value. The question that requires answering, therefore, is why there should be this marked irregularity in the pulsations associated with these stars. Before coming to this, let us first see how these stars reached their present position on the Hertzsprung–Russell diagram. Some of these variables are found within the globular clusters, and on the evolutionary tracks for these clusters they are located near point C in Fig. 13, this being the region where the 'popping' of the core occurs and also the point where we might expect the dimensions of an evolving star having a mass between one and ten times that of the Sun to be at a maximum. Now as a star evolves away from the main sequence, expanding in the process, its surface temperature falls. One further feature is that such a star remains a slowly rotating object and consequently there is still little, if any, mixing of the constituents in its interior.

All of this seriously affects the nuclear reactions that can take place within the star during this particular phase of its evolution. Certainly, as we have seen, the proton-proton chain reaction is predominant in stars such as the Sun, while stars that are somewhat more massive have temperatures in their interiors sufficiently high for the carbon-nitrogen cycle to provide energy. But these are all main-sequence stars and it is important to realize that the nuclear processes which were described in detail at the beginning of this chapter, processes which result in the conversion of hydrogen into helium within the core, are the dominant reactions only so long as the star evolves slowly up the main sequence. Once the amount of helium in the core reaches a critical value and brings about evolution away from the main sequence, the situation changes appreciably.

As may be imagined, the low temperatures prevailing in the long-period variables allow neither the proton-proton chain nor the carbon-nitrogen cycle to go on. Within the past two or three decades, physicists have investigated several other nuclear reactions that can take place at various temperatures and it seems very possible that certain of these reactions provide energy at temperatures lower than in the centre of the Sun. The simplest of these, and the one that is now believed to be operative in the red giants of the long-period variable class, involves the collision of a proton and a deuteron and can occur at temperatures as low as $500,000°K$:

$$^1H_1 + {^2H_1} = {^3He_2}. \tag{14}$$

At somewhat higher temperatures, between $1,500,000$ and $2,000,000°K$, reactions involving lithium, the next higher element to helium, begin, such as:

$$^7Li_3 + {^1H_1} = 2{^4He_2}, \tag{15}$$
$$^6Li_3 + {^1H_1} = {^4He_2} + {^3He_2}, \tag{16}$$

and it may be that in the centres of these red-giant stars both of these reactions are going on, with the former predominating. As very little mixing of these gases occurs, we may postulate that the degree to which each reaction contributes to the total energy production in the stellar interior is not constant, leading to a continuous, but somewhat irregular, production of energy; this has an effect on the induced pulsations, resulting in the wide variations seen in the light curves. In spite of the very large amplitudes of these

118

variables, the range of temperatures encountered within their atmospheres is very small, changing by only a few hundred degrees throughout each cycle.

Shock-Wave Phenomena

There is now a growing amount of evidence for the presence of supersonic shock waves within the extended atmospheres of these giant stars, derived mainly from spectroscopic observations. When we examine the spectrum of a typical long-period variable we find that both the bright emission lines and the absorption lines show a Doppler shift, which indicates that the atmospheric material is moving continuously towards and away from us. In the Cepheids and RR Lyrae stars, the radial-velocity curves are quite regular, but those found for the long-period variables do not show this regularity. Nevertheless, some generalities may be drawn from the evidence available although, as may be expected considering the wide range of periods, amplitudes and spectral types found among these stars, some of these features tend to be rather vaguely defined.

In all of the stars investigated so far, however, the radical velocities of the absorption lines are systematically higher than those derived from the emission lines. This is almost certainly due to the fact that the absorption lines originate much deeper within the atmospheric levels than the emission lines, any shock waves that develop do so at these deeper levels and the velocity of such wave fronts will be greater here before they expend much of their energy in moving outward. As far as the emission lines are concerned, the radial velocities of those due to metals are slightly higher than those of the Balmer series of hydrogen, and there is some evidence, not yet fully substantiated, that even among the individual Balmer lines we find different velocities. All of this evidence is consistent with the idea of shock waves forming deep within the atmosphere of the star.

From the variations in velocity that are found it appears that the speed of the shock waves is of the order of only 10 kilometres per second but, since the atmospheres of these stars are both extremely tenuous and cool (the surface temperatures usually lie between 1,800 and 2,700°K), we can calculate the speed of sound in these regions if we assume that they are composed almost entirely of atomic hydrogen. This turns out to be about 4 kilometres per second. Consequently, the shock waves are travelling at supersonic velocities. We may therefore visualize the situation as follows. A shock wave is generated deep within the atmosphere of the star, below the reversing layer, inside the convection zone, and gives an outward motion to the gas. Now once outside the convection zone the density decreases very

119

rapidly so that although energy is being dissipated quickly, the actual velocity-amplitude of the wave increases. The shock front can therefore travel great distances before all of the energy is lost and it comes to rest. As it passes through the neutral hydrogen of the stellar atmosphere, its energy is sufficiently high to ionize the atoms, stripping off some of the electrons, although behind the wave front these free electrons will re-combine with the protons very rapidly to give neutral hydrogen once more. When we compare the radial-velocity and light curves for these stars, it appears quite probable that the maximum brightness coincides with the maximum strength of the shock wave.

To end this section on the long-period and semi-regular variables, we must examine the masses and also the population to which they belong. We have already mentioned that the RR Lyrae variables are found in large numbers in the globular clusters. The Cepheids, however, are almost totally absent. This enigma was resolved when Baade showed that there are two different stellar populations: those like the Sun, which are found in the arms of the spiral galaxies, known as Population I stars, and those found in the globular clusters, the elliptical galaxies, and the nuclei of the spirals, and between the arms of the spiral galaxies, including our own, which are designated the Population II stars (see Plate X). The Population II stars are those which were first formed out of the primal gas of the galaxies and, besides being the oldest stars, are very deficient in metals. This we would expect since the appearance of the heavier elements within the galaxies is a comparatively recent phenomenon, having come about, as we shall see later, by their synthesis within giant Population II stars which subsequently turned into supernovae, scattering these elements throughout the interstellar gas within the galaxies.

Those variables which are found within the globular clusters, particularly in 47 Toucanae, have mean periods at the lower end of the overall range found for the long-period stars, generally less than 250 days, and also possess high spatial velocities. Quite clearly, since they are members of these clusters, they are Population II stars with masses similar to that of the Sun, their ages being approximately 15,000 million years.

The long-period variables which we find within our own Galaxy, in contrast to those just mentioned, usually have longer mean periods and somewhat smaller spatial velocities. The reason for this is that they are mainly Population I stars and have masses lying between two and five solar masses. In agreement with their population type, they are younger stars with ages around 600 million years. There are, of course, the extremes to be found within each population and several variables are known that have characteristics intermediate between the two.

Evolution and Stellar Variability

Having discussed the main classes of variable stars, we are now in a position to make a more complete and detailed study of stellar variability along the various evolutionary tracks of stars of different masses and of the variety of nuclear reactions that predominate at the different stages. In order to do this we must start at the very beginning of a star's career and trace it throughout its life history.

Once a gas cloud commences to shrink, fragmentation takes place sooner or later and stars begin to form, initially as large masses of gas drawn slowly together by gravitational attraction into more or less spherical shapes which may be anything up to a light-year or more in overall diameter.

With the passage of time the protostar will accrete mass from the surrounding gas, and continued contraction will gradually raise the temperature in the centre to somewhere in the region of 400,000 to 500,000°K. Before this temperature is attained, the star will emit radiation only in the infra-red and, as has already been mentioned, it is interesting to note that such infra-red objects have recently been discovered, notably in the Orion nebula where star formation is known to be in progress.

Depending upon the mass of the protostar, the resulting object will, at one end of the scale, be a T Orionis variable with a fairly large mass and quite a high surface temperature or, at the other end of the range, a T Tauri star with low mass and temperature. Such stars lie to the right of the main sequence, moving on to it in a period of a few million years and joining it at a point dependent upon their mass.

Once the star is on the main sequence, the temperatures within the core are sufficiently high for either the proton-proton chain reaction to begin, as in the case of the Sun, or, for those stars with slightly higher masses and luminosities, for the carbon-nitrogen cycle to become the dominant nuclear process. Now the proton-proton chain reaction does not use up hydrogen at a rapid rate and this phase of stellar evolution is a slow process; the Sun, for example, has taken some 4 million years to reach its present state as a G0-type star. Once a core of helium has formed, with negligible mixing taking place, the central temperature rises, mainly because of contraction, until the carbon-nitrogen cycle becomes the dominant source of energy, even in stars such as the Sun. Now this series of nuclear reactions consumes hydrogen at a much higher rate than the proton-

121

proton chain sequence and consequently the star, which now has an F-type spectrum, moves quite quickly up the main sequence.

When about 10 per cent of the star's mass resides in the helium core, the star commences its evolution away from the main sequence, this being accompanied by expansion and cooling of the outer regions until the red-giant phase is reached. With the lower temperatures prevailing in much of the mass of these stars, only the simplest of nuclear reactions can be maintained, this being, as we have seen in equation (14), the interaction of a proton with a deuteron. Here we are in the realm of the long-period, semi-regular and irregular variables, and it is probably significant that virtually all of the red-giant stars are variable to some degree.

Following the 'popping' of the core, the star now begins its evolution back in the direction of the main sequence, contracting in the process. As the central temperature rises once more, further sequences of nuclear reactions can begin, Lithium becomes reactive at a temperature of around 1,700,000°K, again transmuting hydrogen into helium according to equations (15) and (16). By the time the star enters either the Cepheid zone or the RR Lyrae instability gap (depending upon its mass) the internal temperature has reached between 3,000,000 and 9,000,000°K, and at this stage the next two elements, beryllium and boron, can take part in nuclear reactions as follows:

$$^9He_4 + {}^1H_1 = {}^6Li_3 + {}^4He_2 \qquad (17)$$
$$^{11}B_5 + {}^1H_1 = 3\,{}^4He_2 \qquad (18)$$

Once even higher temperatures are attained, of course, the carbon-nitrogen cycle, already described, is capable of being sustained.

Those stars which rotate rapidly enough for some mixing of the constituents to occur follow a very similar evolutionary track but will move even further up the main sequence before turning off to the right. After crossing the main sequence during the later stages of their evolution such stars undergo gravitational contraction, moving into the region of the blue dwarfs, the more slowly rotating stars ending their careers as white dwarfs.

It will be noticed in Fig. 13 that a third evolutionary track is shown, one that does not begin on the main sequence. Stars which follow this track are not main-sequence stars but are extremely bright giant stars of spectral classes O, B, and W. The general opinion among astronomers at the present time is that these stars have a completely different life history from those we have just discussed. Stars of spectral type W are known as Wolf-Rayet stars after the two astronomers who discovered them. They have very peculiar spectra,

122

indicative of extremely high surface temperatures, often as high as 40,000°K, and it is believed that when initially formed they consist of a relatively dense core surrounded by a diffuse and extended atmosphere. The hot central core radiates strongly in the ultra-violet, whereas the surrounding gas is in the form of a diffuse nebula (the Wolf–Rayet stars show bright emission lines in their spectra that are similar to those of the gaseous nebulae). Owing to their large masses and high temperatures, such stars evolve very quickly indeed, moving almost horizontally across the top of the Hertzsprung–Russell diagram into the region of the red giants, using up their reserves of hydrogen extremely rapidly. Only when the red-giant stage has been reached and contraction begins will the carbon-nitrogen cycle set in, bringing the stars back on to the main sequence, where they move up towards the region of the supergiants. With the depletion of the last stores of hydrogen, such stars then move quickly down into the blue-dwarf stage, their entire lifetime lasting only about a tenth of that of the Sun. The Wolf–Rayet stars are therefore not only very brilliant objects, but also rather short-lived.

The one important group of stars not mentioned so far in this discussion are the novae and supernovae. Since the reactions taking place in their interiors are of such a complex nature, they deserve a chapter to themselves.

5. Exploding Stars

In the previous chapter we discussed the evolution of stars from their initial condensation out of the large gas and dust clouds to their final stages as white or red dwarfs on the slow decline to eventual extinction, and saw how the various evolutionary phases are intimately associated with the nuclear processes taking place in their interiors. Certain of these reactions lead to unstable conditions inside the star, resulting in short-term changes in the energy output which in turn give rise to various classes of variable star. The instabilities which have been considered so far, however, are comparatively mild and on a relatively small scale. Those with which we shall now be dealing are of a much more violent nature, sometimes resulting in the almost complete disruption of the star itself.

We are so adapted to the apparent tranquillity of events taking place within the solar system that when we move out among the stars of the Galaxy as a whole, and even more when we pass beyond its confines into the furthermost depths of space, it comes as something of a surprise to find a picture of almost indescribable violence. The thermonuclear processes that maintain the energy output of the Sun are sufficiently gentle to permit life to exist on Earth. An extremely small variation in the Sun's output of light and heat would result in the total extinction of life on this planet.

The Novae

Before discussing the various reactions that lead to a typical nova explosion we must first consider some of the general features of these stars. On the observational side, novae have been observed for several centuries and, like the comets and eclipses of the Sun, were regarded with a mixture of awe and superstition before their true nature became known. The term 'nova' means quite literally a new star and was given to these objects before refined techniques and accurate star charts giving the positions of very faint stars were available to astronomers. Such charts showing that an extremely faint object existed in the same place as the nova before the actual flare-up occurred.

In general, a nova will brighten from obscurity without any previous warning, increasing in brightness with great rapidity, often

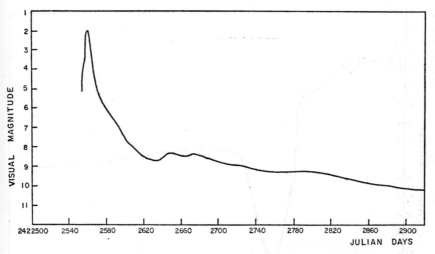

Fig. 25 Light curve of Nova Cygni, 1920, a typical fast nova.

rivalling the brightest stars when at maximum. Its stay at maximum is, however, usually quite short, and a gradual decline sets in with the star eventually fading to something like its initial magnitude. Even after its return to minimum, irregular but small variations in its light often continue, sometimes for several decades. Nova Persei, for example, which attained first magnitude in 1901, exhibits flare-like outbursts to this day, these normally amounting to about a magnitude in amplitude.

Light Variations of the Novae

The vast majority of these stars are characterized by a rapid rise to maximum brightness followed by a slower decline. When we examine their overall changes in brilliance we find differences of detail and it often proves simpler to treat each star individually rather than to group them together into one single class as may be done with most other types of variable star. In spite of these differences (which may sometimes be quite pronounced), it is possible to divide the novae into three broad groups according to the shape of the light curve during the initial rise to maximum and the subsequent behaviour during the first stage of the decline to minimum.

The rapid novae have an extremely steep rise to maximum, often attaining their peak brilliance in a single day, but a sharp decline soon sets in. A purely arbitrary criterion for placing a nova in this

125

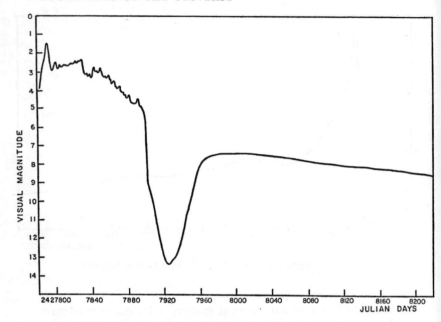

Fig. 26 Light curve of Nova DQ Herculis, 1934, a typical slow nova.

particular class is that it should fade by three magnitudes in a hundred days or less after maximum. After a time the rate of fading slows and the ensuing decline to minimum may then be either smooth (Nova Cygni 1920, Nova Puppis 1942) or pronouncedly irregular with marked fluctuations (Nova Persei 1901, Nova V368 Aquilae 1936). The light curve of Nova Cygni 1920 (Fig. 25) may be compared with that of Nova DQ Herculis 1934, which is a typical slow nova (Figs. 25, 26).

The second group of novae, known generally as the slow novae, show a much more gradual development throughout most of their light curve. The initial rise takes several days, sometimes a fortnight or more, and the star remains around its maximum brightness for several weeks or even months before it begins to decline appreciably. Such stars fade quite slowly at first, the initial drop of three magnitudes taking place over more than a hundred days, and subsidiary maxima are quite common during this phase. Once fading begins in earnest, however, they decline rather quickly to something like their former magnitude (Fig. 26).

126

In contrast to the large amplitudes of the rapid novae, which are of the order of 12 to 14 magnitudes, those of the slow novae are usually smaller. Nova Delphini, discovered by Alcock in 1967, has proved to be one of the most interesting of all these objects. Fortunately for astronomers, not only was this particular star photographed some years before the outburst, but it was also observed spectroscopically when it was a twelfth-magnitude star having an O9-type spectrum. At maximum, this star reached fifth magnitude, its amplitude therefore being only seven magnitudes, which is small for a typical nova. Indeed, since its rise, the behaviour of this star has been far from typical; it remained at maximum for over a year before commencing to fade slowly. Observations made by both the American Association of Variable Star Observers and the Variable Star Section of the British Astronomical Association have confirmed that several individual outbursts occurred following the initial rise, and spectroscopic studies have also revealed the presence of expanding shells of gas thrown off during each outburst.

Once minimum has been reached, the slow novae still continue to show a wide diversity of behaviour. Some, such as Nova DQ Herculis 1934, rise fairly quickly to a secondary maximum, fading smoothly thereafter to minimum. Others, for example Nova Aquilae 1936 I and Nova Pictoris 1925, do not exhibit this secondary brightening and show only small, irregular fluctuations rather like those of certain fast novae.

Finally, there is a very small group of stars which may be described as the ultra-slow novae. Their light curves are very similar to those of the preceding class, but their maxima extend not over the course of a few weeks or months but over several years. Even the inevitable decline proceeds with extreme slowness.

The most spectacular of this small group of stars is undoubtedly η Carinae, which was probably first observed by Halley in 1677 when it apparently varied between second and fourth magnitude in a completely irregular manner. Ptolemy did not make any record of this star in the *Almagest*, which would suggest that it was not as bright as fourth magnitude at that time (circa A.D. 140). In 1810, however, η Carinae began an unparalleled series of brightenings. Over the next decade, it climbed slowly to magnitude 2·0 where it remained at essentially the same brightness until about 1840 when it began to rise again, reaching about magnitude −0·7 in 1843. A slow fading then set in and by 1870 the star had reached seventh magnitude, where it has remained ever since, although even now its brightness is far from constant and fluctuations still occur (Fig. 27).

127

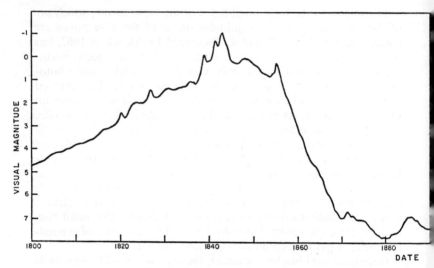

Fig. 27 Light curve of η Carinae.

Spectroscopic and photographic studies of this remarkable object suggest that here we have a huge central star, a supergiant of spectral class cF5 with a surface temperature of about 7,500°K, surrounded by a very extensive envelope or corona, around which has formed an even more extensive shell of gas that is continually being added to by matter ejected from the star itself. Both theory and observation indicate that this star lies far from the main sequence on the Hertzsprung–Russell diagram, but at present it is not possible to say with certainty whether it is evolving towards or away from the main sequence. Its mass has been estimated as between fifty and eighty times that of the Sun and without doubt this means that its evolution will be extremely rapid. The suggestion put forward by Burbidge that this star has lost most of its available hydrogen, either by nuclear reactions within the core or by ejection into the surrounding shell, is probably correct. This, in itself, would indicate that η Carinae is evolving away from the main sequence and has reached a point of instability similar to that of the ordinary novae but on a somewhat milder scale.

Astronomers are now able to put forward certain evolutionary models for stars as massive as this, on the basis of which it is believed that, in those stars with masses less than sixty-five times that of the

128

Sun, any induced pulsations will be stable, whereas above this limit there will be an increasing instability very like that now found in this star.

Burbidge has also suggested that these pulsations will eventually become so marked that a supernova explosion will result. Whether this will indeed be the case, only time will tell. From purely theoretical considerations, however, it does not appear impossible.

Two further members of this class are worthy of mention here. RT Serpentis was not known before 1909, and photographs taken of the region prior to this time indicate that the star was then fainter than sixteenth magnitude. During 1909, however, it began to rise slowly and erratically until it reached magnitude 11·7 this very gradual increase in brightness being maintained for the next ten years. It attained magnitude 10·5 in 1919, remaining there until 1923; it then commenced fading slowly to just below thirteenth magnitude where it has stayed ever since.

The other star appeared in 1939 in the constellation Orion and has been designated FU Orionis. Its light variations were very similar to those of RT Serpentis and its spectrum, like that of η Carniae, is of type cF5, indicating a surface temperature of about 7,500°K. That this is substantially lower than that found for any of the fast or slow novae is probably significant, suggesting that the ultra-slow novae, where the nova-like activity is spread over several years instead of months, have in general much lower surface temperatures.

Spectra of the Novae

The visual characteristics of the novae have been known for several centuries. The reason for their peculiar behaviour has been known for only a few decades and even today there are certain important questions to which answers have yet to be found. During the last century, several theories were advanced to account for the rapid brightening of these stars and their subsequent behaviour. The head-on collision of two dark stars would certainly produce a rapid increase in luminosity and would also account fairly well for many of the features of the light curve. Owing to the utter emptiness of interstellar space within the Galaxy, however, such collisions would be extremely rare events and could not possibly account for the relatively large number of novae that have been observed.

The passage of a dark star through a dense cloud of dust might also produce light changes which would fit the observed variations in brightness, the outer layers of the star being heated to incandescence by frictional and ionization effects. Both very faint stars of low intrinsic luminosity and clouds of dust do, in fact, exist within

I

the Galaxy but several novae have been found in regions of the sky where there is no evidence at all for the existence of dark, obscuring matter, thereby disposing of this particular theory. It was also thought that the rapid increase in luminosity might be due to a large swarm of massive meteorites plunging into the atmosphere of a dim star, drawn to it by gravitational forces. This idea was found to be untenable when calculation showed that the total mass of such a meteorite swarm would have to be impossibly high to produce the observed effect.

The explanation now universally accepted, that of an explosive condition in the outer layers of the star, was provided by the spectro-scope. Although there are wide variations in the light curves of these stars, their spectroscopic characteristics are very similar and show certain well-defined changes following the initial outburst. It will be readily appreciated that, owing to the extreme rapidity of the rise to maximum and the fact that we have no warning whatever of when this will occur, it is only on very rare occasions that any spectra have been obtained during this phase. Almost invariably, the nova is near its peak brightness by the time we are aware of it. The pre-outburst spectrum has been obtained in a small number of cases, where the star has been not too faint, from patrol plates of the area, often taken several years before the star turned nova. Normally, such stars are of Class B or A, but sometimes they may be as early as late-type O. As we have seen, this is the case for Nova Delphini 1967 which was a magnitude 11·9 star with an O9-type spectrum before its recent outburst. The behaviour of this particular nova following its discovery indicates that it may belong to the ultra-slow class, although it is a little early yet to be absolutely certain of this. It has proved to be, without doubt, one of the most unusual novae of the present century (Fig. 28).

At maximum brightness, the nova spectrum is quite characteristic, consisting of a bright continuum with the maximum intensity in the far ultra-violent. Very broad emission and absorption lines are superimposed upon this continuum, these being due to hydrogen; they invariably exhibit a Doppler shift to the violet, showing that the source of these lines is approaching us. From the magnitude of the violet shift it is possible to calculate the velocity of approach of these expelled gases, a typical value being 2,000 kilometres per second. As the brightness of the nova declines, the spectral lines broaden and the underlying continuum fades. For a time the hydro-gen emission lines increase in intensity, then decrease and, as the light continues to fade, bright bands appear in the nova spectrum, these being due to atoms having a very high excitation potential.

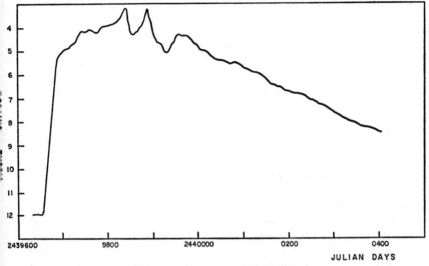

Fig. 28 Light curve of Nova Delphini, 1967.

The temperature as calculated from the spectrum at this stage is of the order of 35,000°K.

Though this is a very general description of the various changes which take place in the spectrum of a nova after the initial maximum, there are naturally some variations from the sequence described. Certain of the stages are sometimes missing altogether and individual stars show other peculiarities. The evidence for a rapidly expanding cloud of gas around a nova is confirmed in many cases, both visually and photographically, once the light of the star has diminished somewhat, when this nebulosity may actually be seen.

The Nova Explosion

Having established that the basic cause of the nova phenomenon is an explosion, we are now in a position to examine in detail the various physical processes that have been suggested to account for this spectacular behaviour. From what has already been said, it is clear that we must look for the origin of the outburst within the star itself. To begin with, what are the facts we can obtain merely by observation?

First of all, there is a tremendous liberation of energy from the

131

outer layers of the star which occurs in a very short time, resulting not only in a sharp rise in temperature but also in the expulsion of a certain proportion of the atmospheric hydrogen with high velocities. Secondly, the total amount of material actually lost by the star during the explosion is quite small, in spite of the large amplitudes of many novae, being not more than about 0·01 per cent of its mass, much of the ejected material eventually cooling down and being drawn back into the star.

This would suggest that the nova phenomenon is a repetitive process, and calculation shows that a star may undergo a thousand such outbursts before it loses all of its excess hydrogen by this means. At first sight this may appear to be a contradiction, since most novae have only been observed to erupt once. However, the interval between outbursts is measured in hundreds or thousands of years, this apparent contradiction disappears. Besides, this view receives additional support from two different sources. To begin with, we can calculate approximately how many novae should appear annually within the Galaxy, on the assumption that there are about 100 million stars which are evolving along a track similar to that shown in Fig. 14, the evolutionary scale of such stars being of the order of 5,000 million years. The result turns out to be about twenty-five a year, which is in excellent agreement with the number observed. Secondly, we might expect to find some stars which undergo the nova explosion at more frequent intervals. Such stars, the recurrent novae, are indeed known and several, for example T Coronae Borealis, RS Ophiuchi, and T Pyxidis, have been comprehensively observed for over a century. It is perhaps significant that the recurrent novae have amplitudes smaller than those of the ordinary novae.

One further observation which must be taken into account when putting forward any theory of the nova phenomenon is that of the spatial distribution of these stars. It has been known for a long time that the great majority of novae occur along the boundaries of the Milky Way. This is not merely because there are far more stars lying in this particular direction but because the novae belong almost exclusively to Baade's Population I stars in the galactic disc. These are the relatively young stars like the Sun, which contain a relatively high proportion of metals. Admittedly, a few novae have been discovered in one or two of the elliptical galaxies and one, T Scorpii, in the globular cluster M 80; these almost certainly belong to the Population II stars, but such novae are in the minority. Later in the chapter, we shall see how this particular observational fact has had a very important bearing upon one theory of the nova phenomenon.

The problem now is to determine just how the explosive condition arises within the star. Even today we are not certain of the precise course, although we know a great deal of the various physical processes that go on in these stars. In the previous chapter we saw how, during the evolution of stars with masses very similar to that of the Sun, there comes a time when the material in the core becomes degenerate, resulting in an explosion which, since it is held in by the weight of the overlying layers of gas, merely has the effect of 'popping' the core and reversing the evolution of the star back towards the main sequence, We saw there, too, that degenerate matter has the peculiar property of not being self-governing but that, once the balance of energy production is upset, the unbalance goes out of control in a very short space of time. The 'popping' of the core, for example, probably occurs within minutes.

We also saw how an energy-generating skin is formed around the core of helium inside a star and, since there is very little mixing of hydrogen and helium, this energy-generating layer gradually works its way out towards the surface of the star as more and more helium is produced from hydrogen. Now just how far can this reaction zone progress towards the surface? Can it, for instance, work its way completely through the material of the star until all of the available hydrogen is used up? A little reflection will tell us that this cannot possibly happen since the temperature of this skin and the energy produced within it are so great that the surface area of the star is just not large enough for all of this energy to be radiated into space. Here we come up against a grave problem, for both Greenstein and Bidelman have observed stars that are apparently burning their hydrogen all the way to the surface.

Either our theory breaks down completely at this stage or there is some means whereby the last of the star's hydrogen may be got rid of without the energy-generating skin's actually working its way out to the surface. Now there are ways by which this may be done. Suppose we modify our earlier assumption that no mixing of hydrogen and helium occurs and accept that, when the energy-generating skin comes close to the stellar surface, convection currents within these outer levels allow some interchange of these gases to take place. We already know there is a convection zone just below the photosphere and this would mean that during the final stages of hydrogen consumption the nuclear reactions would occur in a relatively thick zone compared with the depth of the energy-generating skin and on a somewhat milder scale.

A second, and more probable method, of explaining these helium stars is to say that at some time in the past they lost their envelopes, for example in the form of planetary nebulae, thereby exposing the burnt-out helium core.

There is also another alternative: simply that a large part of the outer layer of hydrogen becomes degenerate! We then have a situation analogous to the 'popping' of the core, but with one very significant difference. The resulting explosion now takes place on the surface of the star where its effects are readily discernible. The tremendous flood of energy released in the explosion can be calculated fairly easily if we make some assumptions regarding the actual amount of hydrogen involved, this energy being converted either into motion of the expelled gases or into heat. This is just the situation we find in the case of the novae, and the figures we obtain are in agreement with observation. There is, however, one particular piece of observational evidence that the theory does not explain, and it is this which we must now consider, since it leads us to an alternative theory of the nova phenomenon.

Binary Systems among the Novae

In 1954, Walker discovered that Nova DQ Herculis 1934 is a close binary system consisting of two small stars, one a red dwarf and the other a much hotter white dwarf. Since that time several other old novae have been investigated, particularly by Kraft, and duplicity has been proved for a very large proportion of them. One of the most remarkable features of these old novae is that they all appear to consist of a blue star and a cooler, red companion; when we consider this statistically, the conclusion becomes well-nigh overwhelming that this particular combination among the novae is far from coincidental. In all cases, too, the red companion is losing mass through the inner Lagrangian point, matter that is either captured by the blue component or takes up an orbit around it in the form of a ring of high-temperature hydrogen. Kraft has even gone so far as to suggest that the rather unusual combination of a blue and a red star is a prerequisite for a star's becoming a nova. This, of course, is a point that is not raised by the theory just outlined, where the presence of a binary system of any kind is not a necessity for the nova phenomenon.

If we assume for the moment that the pre-nova state is that of two physically connected stars of the types just mentioned, we find that there are two possible mechanisms by which the nova explosion may be brought about. In both cases the hot, blue companion is presumed to be the seat of the actual explosion. The first possibility is that the blue star is an electron-degenerate configuration, in other words, that it is very similar to that discussed earlier in which the mass of the material is degenerate with only a thin shell of non-degenerate hydrogen forming an outer envelope, the energy-generating skin having worked its way through the mass of the star almost to the surface. This is what we normally find for white dwarfs. Now from observation of several old novae we also find that there is a stream of gas, mainly hydrogen, in these systems which is almost certainly flowing from the cool companion to the hot component. We may therefore envisage a situation in which this stream of gas, at a high temperature, penetrates the thin shell of non-degenerate matter surrounding the core of the blue star and mixes with the degenerate material in the centre. The result of this mixing is reasonably predictable. Inside the helium core the temperature is certainly high enough to start the hydrogen burning. Since the material inside the blue star is degenerate, the result will be a tremendous increase in the reaction rate, and the star, will be unable to cool off by expansion as it would if the matter were not degenerate. Now if the degeneracy can be removed rapidly, the blue star will become a nova.

So far this theory fits several of the observed facts even though, at the moment, we do not fully understand the mechanism whereby the degeneracy can be removed as rapidly as required from observation of a typical nova outburst. This, however, is not the main stumbling block to this theory. It is when we come to examine the behaviour of degenerate matter under these conditions that we encounter difficulties. One property of such matter is that it is an excellent conductor of heat, and it seems unlikely that the outer regions of the degenerate core can be heated up quickly enough to give rise to a nova explosion. If we go further than this and imply that there may be an induction period during which the entire core is heated to a sufficiently high temperature to explode we find that, although this is theoretically possible, the result will be the release of so much energy that the star will become not a nova, but a supernova! Reluctantly, therefore, we must discard this idea in spite of its many attractive points unless future theoretical work on degenerate matter provides some way of getting over this difficulty.

Fortunately, there is one additional feature of close binary systems which may help us out of this dilemma. A large number of these systems are known in which eclipses occur, these being the celebrated Algol variables. Because these are eclipsing systems, we are able to infer most of the physical parameters of such systems from their light curves and spectra, factors such as the individual masses of the components, their diameters, periods of orbital and axial revolution and densities. In this particular instance, it is the periods of orbital and axial revolution that are of interest to us. From a detailed study

135

of many of these stars it has been found that, in certain cases, the period of revolution of the component stars on their axes is not synchronized with the orbital period of the system.

Since the Algol variables are all close binaries, we would expect to find, in the absence of any interfering mechanisms, a very close similarity between these two periods, owing mainly to the powerful gravitational fields existing between the two stars. This is, of course, exactly what we do find in a great many cases, but it is significant that where there is non-synchronization we also find a gas stream present in the system. The effect of the gas stream is to carry angular momentum from the fainter star to the brighter.

Now a similar effect has been found in the old novae, and about thirty years ago Cowling showed that in certain instances where this occurred non-radial oscillations may be induced in a star and, provided that certain conditions are fulfilled, these may come into resonance with the orbital period. Recently, Schatzman has applied this idea to the novae and demonstrated that eventually these resonating non-radial oscillations will build up and lead to the ejection of matter from the blue star. One important point to make here is that this material will be ejected in the form of cones and not as a spherically expanding shell. The importance of this is that in a large number of cases we observe that the gases thrown out by the novae are expelled in two preferred directions. This was particularly the case for Nova DQ Herculis 1934; here the greater portion of the gas was ejected sideways across our line of sight, and when seen in the telescope some time after maximum had passed it appeared as though there were two small clouds of nebulosity on either side of the central star.

This is a very strong point in favour of this theory since neither of the others just described provides any theoretical explanation of this phenomenon. One point against the theory is that the nature of the secondary component does not appear to be critical; certainly it does not stipulate that the star must be a red dwarf which is overflowing its inner Lagrangian surface and ejecting mass through the inner Lagrangian point, as has been observed for all of the old novae which have been shown to be binary systems. This section ends, therefore, on a note of uncertainty which is perhaps inevitable. Far more observational data must be accumulated before we can put forward a theory that is completely satisfactory in accounting for the nova phenomenon. However, there is one further piece of evidence in favour of the binary-star hypothesis which we must now consider.

Novae Among the Population I and II Stars

The point has already been made that by far the greater number of novae have been found close to the boundaries of the Milky Way and that they belong to the Population I stars of the galactic disc, which is a relatively flattened region of stars. Very few indeed have been discovered among the much older Population II stars of the galactic halo. Now we can relate this to the number of binary systems found among these two different populations. Among the stars of the galactic disc, the number of binaries is very large compared with those found within the halo.

How do we know this? Chiefly because of the numbers of Algol variables which have been discovered within these two regions of the Galaxy. The light variations of these stars are very distinctive and we may therefore use them as a guide to the relative numbers of such binaries in the galactic disc and halo. Now there happens to be a very good reason for this dearth of binaries among the Population II stars. Binary stars will be formed preferentially in regions of high angular momentum, for it is here that we find stars having low spatial velocities but high rates of orbital and axial revolution. These conditions apply very well to the galactic disc whereas the galactic halo has a low angular momentum and the stars contained within it nearly all have small axial velocities.

This is all borne out by observation and if we make a precise calculation of how many novae we would expect to find among the Population II stars of the halo, the basis of the number of binary systems present among these stars, and compare this with the number we would expect among the Population II stars, we find quite good agreement with observation.

The Supernovae

Early in October 1572 a brilliant star appeared in the constellation Cassiopeia which at maximum was the equal of Venus and visible in daylight. Just over a quarter of a century later, in 1604, a similar star flared up in the constellation Ophiuchus, its light variations following a similar course to that of 1572. Both of these remarkable objects were closely studied, the former by Tycho Brahe and the latter by Kepler. During the course of the next three centuries, a few similar stars were discovered, although none were as bright at maximum, and the general opinion was that they all belonged to the same class of object, namely the novae that have just been described.

In August 1885 a new star appeared quite suddenly close to the centre of the spiral galaxy in Andromeda, its brightness being such

137

that it completely transformed the appearance of the nebula, and for a few weeks it was just visible without instrumental aid. By mid-1887 the star had faded to well below sixteenth magnitude and it was universally assumed to have been just another nova, possibly within the environs of our own Galaxy and merely seen projected against the field of the Andromeda spiral. Even those astronomers who were of the opinion that the star might actually have been within the nebula itself did not consider it of any special importance since at that time neither the nature nor the distance of the Andromeda nebula was known and it was thought that all of the nebulae lay in, or very close to, our own Galaxy.

By 1917, however, the 100-inch reflector had come into operation and several excellent photographs of the nebula had been obtained, particularly by Ritchey, who discovered numerous faint novae on the plates. Within the next fifteen years, more than a hundred of these novae had been located in the Andromeda nebula and, from a study of their absolute magnitudes at maximum, it was soon proved that they were similar to the ordinary novae which frequently appear within our own Galaxy. Moreover, by this time the controversy over the nature and distances of the nebulae was being resolved, chiefly by means of the 100-inch telescope, and it was realized that they were, in reality extra-galactic objects at tremendous distances, each containing hundreds of thousands of millions of stars. The final proof was provided by the measurement of their distances by means of the period-luminosity law for the various types of variable star found in the nearer ones, particularly the Cepheids and long-period variables. Once it was recognized that the Andromdea spiral lies at a distance of 2,200,000 light-years, it became clear that the star of 1885, whose brightness at maximum was almost equal to the integrated light of the whole nebula, was in a very different class from the normal novae.

On the assumption that this star did actually lie within the boundaries of the nebula and was not simply a foreground star in our own Galaxy, the fact that at maximum its apparent magnitude was about 5·8 means that its luminosity must have been equal to that of 1,000 million Suns and its absolute magnitude about −17, which is far greater than that of any normal nova.

Naturally, as soon as this fact was recognized, astronomers began an intensive search for other nebulae, a search that soon resulted in the discovery of several more supernovae, for example that shown in Plate XI. A comprehensive study of the absolute magnitudes of these stars shows that they lie in the range −13 to −17 when at maximum, as compared with −8 to −12 for the ordinary novae. Clearly, the amount of energy released by a single supernova is something on a truly stupendous scale, equalled only by that of the recently discovered quasars.

Light and Spectroscopic Changes in the Supernovae

It is convenient here to consider the light and spectroscopic variations of the supernovae together, since they keep in step remarkably well throughout the entire light cycle and the light curves found among the supernovae are far less diverse than those of the novae. Indeed, we can often tell from the spectrum of a supernova just when maximum brightness occurred even if the actual maximum itself passed unobserved.

Again, the emission and absorption lines in the spectra of these stars provide proof of the explosive ejection of hydrogen from the star during the course of the explosion, and it is significant that the velocity of the expelled gases is virtually the same as that found for the normal novae. This at once suggests that in both cases the explosion is brought about by the uncontrolled liberation of energy by nuclear reactions going on within the star. Why then do we find such a wide difference in their amplitudes and therefore in the total energy released? The answer lies in the amount of stellar material involved in the detonation. In the case of the novae, we saw how this amounts to only about 0·0001 solar mass. The supernovae, on the other hand, throw off at least one solar mass of gas and are therefore some 10,000 times brighter at maximum.

The next question concerning the supernovae is whether we can divide them into two or more classes as we did for the novae. We have already said that in their light variations they show far less diversity than the novae. In spite of this, Baade has been able to show the existence of two main types, basing his classification upon the absolute magnitudes, the general shape of the light curve around maximum brightness, and the spectra. A further two types have been added recently, but these show only slight differences from those originally suggested by Baade.

The Type I supernovae have absolute magnitudes lying between −14 and −17 at the upper end of the scale, their light curves being characterized by a sharply-peaked maximum, followed by a rapid decline for a period before the rate of fading slows and minimum is reached. The supernova of 1885, designated S Andromedae, appears to have been of this type. From their spectra, taken shortly after maximum, we find that the expelled gases are remarkably deficient in hydrogen and the mass of gas ejected is almost invariably less than that thrown out by the Type II supernovae. The colour of these stars around maximum is usually reddish although this soon fades through yellow to a dull leaden colour. It is now believed that these are stars of Population II, very old stars which were formed from the primal gas of the galaxies, and their masses lie between 1·5 and 2·0 times that of the Sun. The radial velocity of the ejected gases is quite low, averaging about 1,700 kilometres per second.

The Type II supernovae, on the other hand, are found only within the spiral arms of the galaxies and clearly belong to the Population I stars, similar to the Sun in many respects but much more massive. Indeed, their masses are often as high as thirty times that of the Sun. It seems quite probable that such stars are still closely associated with the vast gas and dust clouds from which they originally condensed. The light curves of the Type II supernovae are somewhat different from those of Type I. For example, they remain at maximum brightness for a much longer period and their subsequent fading seems to follow a smooth exponential curve. At maximum, too, their light is distinctly blue and, in contrast to the former class, the mass of the gaseous envelope thrown out by the explosion can be as high as five solar masses, this gas being very rich in hydrogen. The absolute magnitudes at maximum are generally lower than those of the supernovae of Type I, ranging from −12 to −13·5, and their overall amplitudes are somewhat smaller. The outward-moving gases, however, may have velocities up to 5,000 kilometres per second.

So much for the various characteristics of the two main classes of supernovae. Before continuing with the discussion of the physical and nuclear reactions which may lead to the actual supernova detonation, we must first examine the frequency with which supernovae occur in the galaxies. Although the extensive photographic search carried out once the true nature of these stars was recognized revealed several supernovae, it must not be imagined that this implies that they are of frequent occurence. We must remember that the survey took in several hundred galaxies and we now know that a supernova explosion is a much rarer event than that of a normal nova. On the average, there is one supernova per galaxy ever 500 years.

This brings up to the question of the number of supernovae which have appeared within our own Galaxy in historical times. Quite clearly, since they are relatively near stars, they will have been unmistakable objects when at maximum. The two stars of 1572 and 1604 were clearly supernovae and from the records we have it seems that both were members of Type I (Fig. 29). A third supernova appeared in 1054 in the constellation Taurus and, although it appears to have passed completely unnoticed in Europe, there are excellent descriptions of it among old Chinese chronicles. Unlike the two later supernovae, that of 1054 has left a visible testimony of the original detonation – the Crab nebula – (see Plate XII) which is still expanding at the rate of about 1,000 kilometres per second.

Apart from being the remains of a supernova explosion, and also a source of intense radio emission, the Crab nebula has recently become important since that discovery that the stellar remnant of the supernova is a pulsar. More will said be about this remarkable object, however, in a later chapter.

A fourth star that is thought by some astronomers to have been a supernova appeared in the constellation Puppis in 1006. There are unfortunately only scanty records of this star, which lies well to the south of the galactic equator, and from the position given it appears probable that this was the star κ Puppis which turned nova. Since this star is quite bright, even at minimum, an amplitude of 8 to 11 magnitudes, such as we find in the normal novae, would result in the star's being exceptionally bright at maximum and resembling a supernova in this respect.

The Physics of the Supernova Phenomenon

In the previous chapter we saw how the evolutionary path followed by these stars was governed by Chandrasekhar's limit – that it held only for those stars in which the helium core had a mass less than 1·44 times that of the Sun. As such a star enters upon its final stages of evolution, the degenerate matter within the core can maintain the balance of pressure and allow the star to cool off slowly towards the final white-dwarf stage. Where the mass of the helium core exceeds this critical limit, evolution takes a completely different and ultimately catastrophic course, as we shall now see. Here degeneracy pressure is no longer the operative force, the pressure balance being maintained instead by high temperature and as a result the star cannot cool off but is forced to continue providing energy within the core by raising the internal temperature. Now there are two ways by which this may be done, either by nuclear reactions within the core (and these will provide energy for only a certain length of time before all of the nuclear fuels become exhausted) or by gravitational contraction. Clearly the latter would appear to be the dominant process as evolution proceeds, but even here we come up against a grave problem, namely how far the star contracts. Theoretically it could continue to do so until it was a very small body indeed, only a few kilometres in diameter. Before this ultimate stage is reached, however, something very dramatic happens: the star literally blows

141

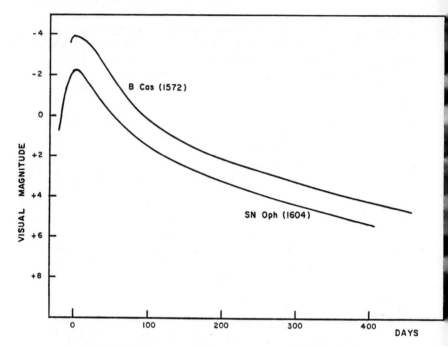

Fig. 29 Light curves of the supernovae of 1572 and 1604.

itself to pieces. Until a year or so ago it was thought that, once the star lost sufficient of its mass to bring it below Chandrasekhar's limit, it would then be able to continue evolving as a white-dwarf star, this being the ultimate remnant of a supernova explosion. Now, with the recent discovery of the pulsars, it seems that this is not the case but that the end product of a supernova explosion is a neutron configuration with a diameter of only a few kilometres. More will be said about this when we come to discuss the plusars themselves.

During the initial stages of its evolution, a star of large mass behaves in a similar way to normal stars with masses like the Sun. Hydrogen burning begins in the centre with the formation of a helium core. Eventually, contraction of the core sends the central temperature up to around the 100,000,000° mark. At this point the helium, normally an inert material, begins burning, with the formation of oxygen and neon. Over a period of time, of the order of four or five million years, the inner core shrinks slowly, pushing up the internal temperature in this region still further, until neon and oxygen burning commences at around 600,000,000°K. We now have a star in which there are several layers, all with different nuclear reactions going on and with each layer separated from the others by an energy-generating skin.

It is very difficult according to present-day theories to determine whether any mixing of material among these layers occurs, and, if so, to what extent. It does seem quite probable that some mixing takes place, with small amounts of hydrogen penetrating into the helium-burning zone where a number of subsidiary reactions then go on, producing elements of still higher atomic weight. The main reason for believing this to be the case is that there is fairly strong evidence that such elements are manufactured in this way deep within the Population II stars and then scattered throughout the galaxy by means of supernova explosions in such a way that the younger Population I stars which are formed later contain small amounts of these heavier elements in their make-up, elements which it is doubtful that they synthesize for themselves during the ordinary course of their evolution.

Once all of the neon and oxygen have been exhausted, the central regions of the star begin to shrink again, forcing the temperature up to still higher levels at which other reactions are able to take place forming, at a temperature of about 1,500,000,000°K, all of the chemical elements up to calcium. Even this is not the end of the story for, as time goes on, the core is forced to contract still further to provide the necessary output of energy, a requirement which becomes increasingly urgent since now a tremendous amount of energy is lost to the star both by gamma radiation and, in the subsequent stage when the iron group of elements are synthesized, by neutrino emission. By this time the central temperature is well in excess of 2,000,000,000°K and we have already seen that neutrinos take part in only certain special reactions. Even in a star such as this, the vast majority of neutrinos take away energy from the star far more efficiently than the normal process of radiation would do.

A further effect of this extremely high temperature inside the core is that not only do these reactions consume nuclear fuel at an ever-increasing rate, but the shrinkage itself accelerates drastically. Whereas it was appreciable only over a million years or so during the neon-burning phase, it becomes appreciable in a matter of weeks as the temperature climbs towards the 5,000,000,000°K mark. By now, the silicon group of elements, up to and including calcium with 20 protons in the nucleus, have begun reacting to form a new set of nuclei, those of the iron group of metals – titanium, vanadium,

143

chromium, manganese, iron, cobalt, nickel, copper and zinc. This is now the beginning of the final stage before the supernova explosion.

To understand the remarkable sequence of events which follows, we must first mention the two types of nuclear reaction that can occur at these extremely high temperatures. Up to now, all of those we have been considering are known as *fusion* reactions since they produce energy by the fusing together of lighter particles (or nuclei) to form heavier atoms. Now even at temperatures as high as 5,000,000,000°K, nuclear reactions can only go on provided they produce or can absorb energy, the former occurring only when nuclei having what is known as a low binding energy per proton (or neutron) produce those having a higher binding energy. Experiments in the laboratory have proved quite conclusively that those nuclei which are heavier than the iron group of elements have lower binding energies than the elements of the iron group themselves. All this means is that the heavier nuclei do not produce energy by the process of fusion. They do, however, liberate energy by the opposite process of fission. These two forms of nuclear reaction are well known in this atomic age. The atomic bomb, based upon the heavy radio-active elements such as uranium and plutonium, liberates energy by the fission of these materials into lighter elements such as barium; the hydrogen bomb, on the other hand, does exactly the same thing by the fusion of hydrogen nuclei into heavier elements.

What happens, then, in the core of the star once the temperature there reaches 5,000,000,000°K is that fission occurs among the iron group, with the complete conversion back into helium. This reverse process, of course, requires the absorption of all the energy liberated over millions of years during the building up of the iron group from helium throughout all the various stages of nuclear synthesis. How does the star find the necessary energy to do this? Clearly there is only one course left open to it now, that of gravitational contraction. On this occasion, however, shrinkage occurs with extreme rapidity. Gravitational infall of the central regions takes place in the space of a few seconds. With the pressure balance upset so abruptly, the outer regions also collapse, raising the temperature still further.

However, it is not this collapse, catastrophic as it is, which produces the explosion but the fact that these outer levels, where hydrogen and helium burning is still in progress, have a very high reserve of energy in them. With the sudden heating of these gases to extremely high temperatures, the reaction rates are increased enormously and it is this sudden release of energy on a large scale

The crater Copernicus. The faint outlines of the rays are clearly shown in this photograph taken with the 200-inch Hale telescope. (*Photograph from the Mount Wilson and Palomar Observatories.*)

The northern portion of the crater Copernicus. The numerous rilles in the floor of this crater and the terraced structure of the ringwall are readily visible. The photograph was taken from a height of sixty-four miles on August 16, 1967 by Lunar Orbiter V. (*National Aeronautics and Space Administration.*)

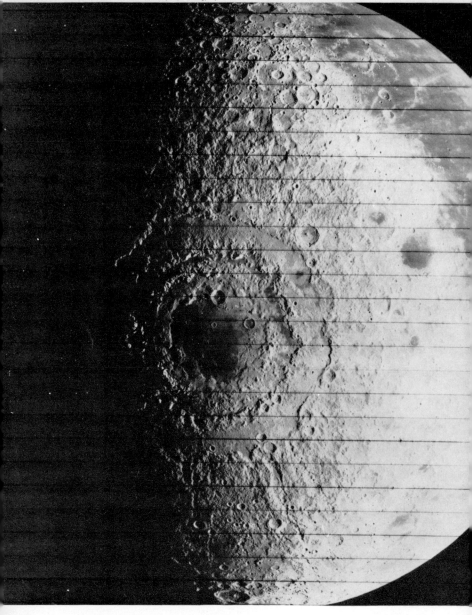

III Mare Orientale. The sharpness of the surrounding mountain range and the well-preserved texture of the crater floor suggest that this is one of the youngest craters on the lunar surface.
(*National Aeronautics and Space Administration.*)

IV The region around Goclenius. One prominent feature of this large
crater is the well-defined rille system crossing the crater floor.
(*National Aeronautics and Space Administration.*)

v The Martian surface. This picture of the south polar cap of Mars was taken by Mariner 7 from an altitude of about 3,300 miles on August 4, 1969. The Martian craters are clearly more eroded than those on the lunar surface.
(*National Aeronautics and Space Administration.*)

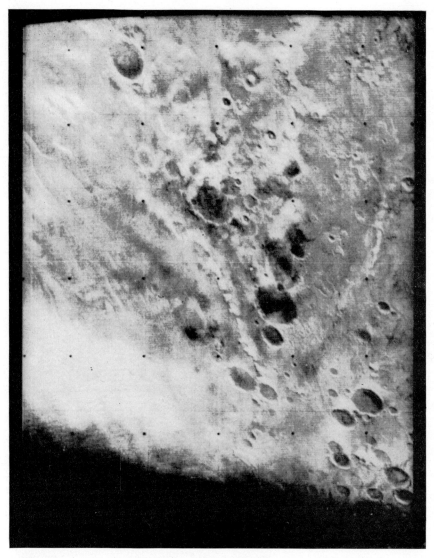

VI The Martian south polar region. The south polar cap, which varies
with the season, is composed either of a layer of solid carbon dioxide a
few feet in thickness or of ice less than an inch in depth.
(*National Aeronautics and Space Administration.*)

VII The planet Jupiter photographed in blue light with the 200-inch Hale telescope. The banded appearance and the Great Red Spot are clearly visible.
(*Photograph from the Mount Wilson and Palomar Observatories.*)

VIII Saturn photographed in blue light with the 200-inch Hale telescope.
(*Photograph from the Mount Wilson and Palomar Observatories.*)

IX The Orion nebula. This great cloud of glowing gas is teeming with young, hot stars, many of which are variable.
(*Photograph from Mount Wilson and Palomar Observatories.*)

STELLAR POPULATIONS I AND II

ANDROMEDA NEBULA photographed in blue light shows giant and super-giant stars of POPULATION I in the spiral arms. The hazy patch at the upper left is composed of unresolved Population II stars.

The very bright, uniformly distributed stars in both pictures are foreground stars belonging in our own Milky Way system

NGC 205, companion of the Andromeda Nebula, photographed in yellow light shows stars of POPULATION II. The brightest stars are red and 100 times fainter than the blue giants of Population I.

x Stellar populations. The left-hand photograph of part of the Andromeda nebula taken in blue light shows the giant and supergiant stars of Baade's Population I within the spiral arms. The right-hand picture of NGC 205, a companion elliptical nebula photographed in yellow light, shows Population II stars, the brightest of which are a hundred times fainter than the blue supergiants of Population I. The very bright, uniformly distributed stars in both pictures are foreground stars belonging to our own Galaxy. (*Photograph from the Mount Wilson and Palomar Observatories.*)

SUPERNOVA IN IC 4182

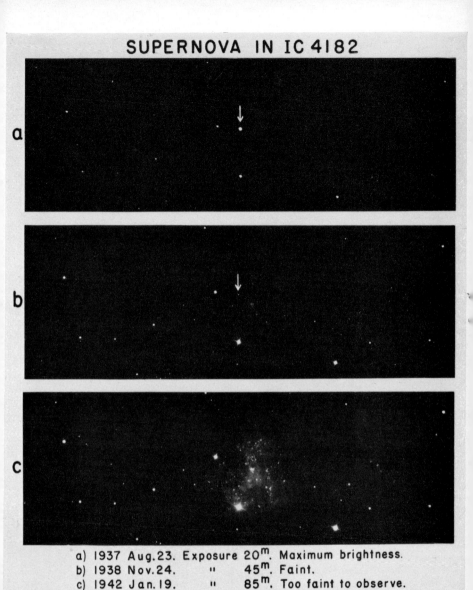

a) 1937 Aug.23. Exposure 20^m. Maximum brightness.
b) 1938 Nov.24.   "   45^m. Faint.
c) 1942 Jan.19.   "   85^m. Too faint to observe.

xi The supernova in the galaxy IC 4182: (a) August 23, 1937 at maximum brightness, 20-minute exposure; (b) November 24, 1938, barely visible, 45-minute exposure; (c) January 19, 1942, too faint to observe, even with an 85-minute exposure.
(*Photograph from the Mount Wilson and Palomar observatories.*)

XII The Crab nebula in Taurus, the remnant of the supernova of 1054,
photographed in red light. The filaments of glowing gas are in
rapid motion; the diameter of this gas cloud is now a little over
30,000,000,000,000 miles.
(*Photograph from the Mount Wilson and Palomar Observatories.*)

XIII The Horsehead nebula in Orion, photographed in red light with the
200-inch Hale telescope. This mass of gas and dust effectively obscures
the stars that lie beyond it.
(*Photograph from the Mount Wilson and Palomar Observatories.*)

xiv The Lagoon nebula, Messier 8, in Sagittarius. Numerous dark globules
may be seen in projection against the bright gaseous background. It
has been suggested that these are stars in the process of formation.
(*Photograph from the Mount Wilson and Palomar Observatories.*)

xv Bright and dark nebulosities in the region of γ Cygni photographed with the 48-inch Schmidt telescope.
(*Photograph from the Mount Wilson and Palomar Observatories.*)

XVI This very peculiar galaxy in Centaurus is a powerful emitter of radio waves. Whether or not it consists of two galaxies in collision is still open to question.

(*Photograph from the Mount Wilson and Palomar Observatories.*)

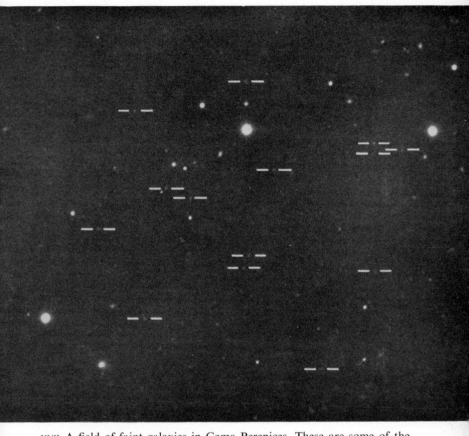

XVII A field of faint galaxies in Coma Berenices. These are some of the most distant objects known.
(*Photograph from the Mount Wilson and Palomar Observatories.*)

RELATION BETWEEN RED-SHIFT AND DISTANCE
FOR EXTRAGALACTIC NEBULAE

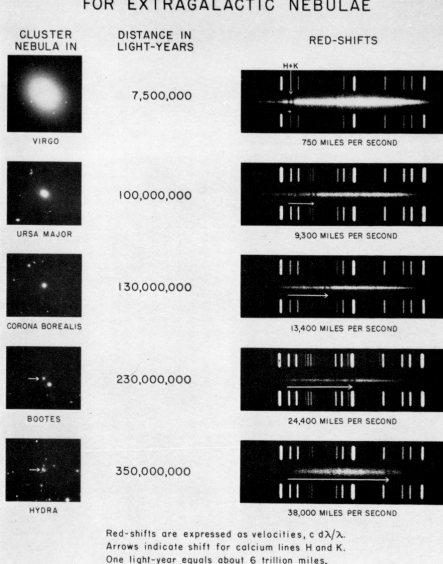

CLUSTER NEBULA IN	DISTANCE IN LIGHT-YEARS	RED-SHIFTS

H+K

VIRGO — 7,500,000 — 750 MILES PER SECOND

URSA MAJOR — 100,000,000 — 9,300 MILES PER SECOND

CORONA BOREALIS — 130,000,000 — 13,400 MILES PER SECOND

BOOTES — 230,000,000 — 24,400 MILES PER SECOND

HYDRA — 350,000,000 — 38,000 MILES PER SECOND

Red-shifts are expressed as velocities, $c \, d\lambda/\lambda$.
Arrows indicate shift for calcium lines H and K.
One light-year equals about 6 trillion miles,
or 6×10^{12} miles

XVIII The red shifts of the H and K lines of calcium in the spectra of these galaxies provide a measurement of their velocities of recession and when plotted against their distances show a linear relation between the two parameters.
(*Photograph from the Mount Wilson and Palomar Observatories.*)

which brings about the final catastrophe. The heated gases are expelled by the detonation, with velocities sometimes in excess of 4,000 kilometres per second.

Earlier in this chapter we saw that there are two main classes of supernovae associated with stars which have widely differing masses and belong to the two different populations. This would suggest that there may be two different mechanisms by which the supernova explosion is brought about.

Recent calculations indicate that the Type II supernovae, the more massive of the two, do not have degenerate cores and that the basic cause of the explosion is the infall of the high-temperature core. Furthermore, we would expect the sequence of events just described to occur preferentially in stars that have a large reserve of hydrogen in their atmospheres, and this again fits in well with what we know of the hydrogen content of the expanding gas clouds surrounding the Type II supernovae.

The next question is whether we can postulate an alternative mechanism for the Type I supernovae, bearing in mind that these are stars with masses between 1·5 and 2·0 times that of the Sun and consequently only normal amounts of hydrogen in their atmospheres. This is a very narrow range of masses, it is true, but it is one which contains a relatively large number of stars, sufficient to account for the observed number of supernovae of this type.

In this case, we shall assume that almost all of the star is composed of degenerate matter before the outburst, with only a small proportion of non-degenerate hydrogen surrounding the core. Such a star will, to all intents and purposes, appear to us as a giant far to the right of the main sequence on the Hertzsprung–Russell diagram. The temperature inside the core can rise as high as 2,000,000°K before the material becomes degenerate at something like twice this temperature. Under normal conditions, a temperature of 2,000,000°K is above the explosion point of the nuclear fuels in the interior of the star, but no explosion occurs since the material is not yet degenerate. This is an important point. For the supernova explosion to take place, two conditions must be satisfied: the temperature must be sufficiently high and at least 90 per cent of the material in the star must be degenerate. As a consequence, the energy-generating skin where hydrogen burning is taking place will almost have reached the stellar surface by the time these two conditions have been satisfied. The detonation is very similar to that already described for the Type II supernovae but with the important exception that very little unburnt hydrogen is now present. The ejected gases will therefore contain a much smaller amount of hydrogen than in the Type II supernovae, again in good agreement with observation.

One apparent discrepancy now arises from what has just been said. The explosive nuclear reactions are those in which oxygen or carbon burning are involved but, curiously, these reactions deliver less energy than the non-explosive hydrogen-burning reaction. Now we have just seen that it is in the Type II supernovae that we have this large mass of hydrogen available for reaction at the moment of the detonation, and we would therefore expect the Type II supernovae to be the brighter of the two at maximum, a conclusion which is in conflict with observation. How can we explain this apparent anomaly?

Referring back to the light curves of the Type I and Type II supernovae, we recall that, although the amplitudes of the former class are generally greater than those of the latter, the Type II supernovae remain at maximum for a longer period and are also distinctly blue at this phase. Here we have the vital clues. The total energy given out during the explosion is spread over a much longer period in the Type II supernovae and, if we integrate the area beneath the light curves from the time of the rise to maximum to the eventual fall to minimum, we find that this is somewhat less in the case of the Type I supernovae. In addition, there seems to be a need to apply quite a large bolometric correction to the Type II stars and clearly much of the energy is coming through in the ultra-violet and does not make itself manifest in the visual light curve.

The problem of the Type I supernovae has recently received detailed study and it is apparent that very complex reactions take place within these stars just prior to the outburst. After the initial explosion, the light curves of these stars appear to indicate exponential decay, as we have seen, and this has led to the suggestion that fission of heavy radio-active elements might be responsible for the greater part of the energy released. These are the transuranic elements, with atomic weights higher than that of uranium. They are very unstable and are not found on Earth, with the possible exception of plutonium; there is some evidence that a very small amount of plutonium may occur naturally. At present, quite a large number of these elements are known, having been produced artificially in the laboratory by bombardment of heavy elements with various lighter nuclei; most of this work has been carried out at Berkeley in California.

The element chiefly involved in the fission process is believed to be an isotope of californium with an atomic weight of 254. This and other similar elements are spontaneously fissionable, with half-lives

146

ranging from a few seconds to about a year. Once the detonation occurs within the core a shock wave of considerable violence passes outward through the external levels of the star, moving through the helium-burning zone in about ten seconds. What happens next takes place very rapidly indeed. Some of the helium nuclei in these outer levels have already been built up into elements such as carbon, and it is thought that these heavier elements can add on more helium nuclei (alpha particles) to form still heavier elements which in turn act as 'seed' nuclei for the ultimate formation of the very heavy radio-active transuranic elements.

Supernova Remnants

It seems obvious that an explosion on the grand scale of a supernova should leave some visible evidence even centuries after the outburst. One such supernova remnant has already been mentioned briefly – the Crab nebula – and others have been found within the Galaxy, although none is as conspicuous. A fortunate characteristic of these supernova remnants is that they are strong emitters of radio waves, which enable them to be picked out fairly readily. Not all radio emitters within the Galaxy are the remains of supernovae, of course, but, provided the time that has elapsed since the outburst is not so long that all of the expanding gas cloud has dissipated into space, we can generally see these faint wisps of gaseous matters, which often possess a characteristic filamentous structure. With the discovery that most, if not all, of the stellar remains of supernova outbursts are pulsars we have a further powerful means of identifying them since the pulsars are easily located by virtue of their radio emission.

The whole question of the radio emission both from the expanding gas cloud and from the tiny stellar remains associated with supernovae is an extremely complicated one, which is still only partly understood, and we shall defer a full discussion of it to a later chapter. We end the present discussion on these stars by noticing that a fundamental feature of fast-moving filaments of gas which interact with the large-scale magnetic field of the Galaxy is that they emit radio waves.

The Dwarf Novae

The exploding stars we have been discussing so far – the novae, recurrent novae and supernovae – are characterized by both large amplitudes and long periods between outbursts. By its very nature, a star can undergo a supernova explosion only once in its entire career. The novae have periods of at least several centuries, possibly

147

many thousands years, whereas the recurrent novae such as T Coronae Borealis and T Pyxidis erupt at intervals of some tens of years.

We must now consider a fairly small class of variable star, the dwarf novae. Such stars have several features in common with the novae and more particularly with the recurrent novae, features such as an abrupt, nova-like increase in brightness at irregular intervals and the fact that all seem to be close binary systems consisting of a white dwarf and a cooler red companion.

The first of these stars to be discovered, U Geminorum, was found by Hind in 1855 when it was at ninth magnitude. Three weeks later it had faded to below thirteenth magnitude and was considered to be merely a faint nova. In March 1856 it was seen again by Pogson and subsequent observation demonstrated quite clearly that it was not a nova but a completely new kind of variable star. Since that time, about 150 of these variables have been discovered; they exhibit rapid, temporary outbursts of light during which they can brighten by as much as five magnitudes in a single night.

Light Variations of the Dwarf Novae

In many ways the light curves of these stars are similar to those of the novae and recurrent novae, the main feature being an abrupt rise to maximum, followed by a slower decline. Their amplitudes are, however, much smaller and the intervals between their maxima far shorter. The prototype, U Geminorum, together with many other members of this class, show two fairly distinct kinds of maximum – long and short – defined by the length of time spent around maximum brightness. Almost invariably the long maxima are brighter than the short.

Although the interval between successive maxima is quite unpredictable, it is possible to obtain mean periods for most of these stars by averaging the intervals over a long period but, although a great deal of statistical work has been carried out on this group of stars, no relationship between amplitude and mean period has yet been discovered.

The brightest and most comprehensively observed of all the dwarf novae is SS Cygni, every maximum of which has been recorded since its discovery by Miss Wells in 1896. From the light curve of this star (Fig. 30) it will be seen that here there are three kinds of maxima, usually termed long, short and anomalous, the last being defined by a more gradual rise to maximum, often with quite pronounced irregularities on both the ascending and descending branches of the light curve.

The individual idiosyncracies of most of the dwarf novae may

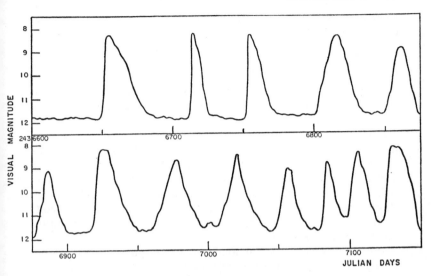

Fig. 30 Light curve of SS Cygni.

be quite marked, but the general pattern behind their light cycles varies little from one star to another. At the beginning of the century, however, one particular star – Z Camelopardalis – was found to differ markedly from the others known at that time. Following certain outbursts, the star did not decline smoothly to minimum but remained at an intermediate brightness for several weeks before the eventual decline to minimum was resumed, after which the normal cycles recommenced. These periods of relative quiescence are known as standstills, and they have been found in a small number of the dwarf novae. Astronomers now divide these stars into two subgroups, the U Geminorum variables, in which only normal behaviour has been observed, and the Z Camelopardalis stars, which exhibit not only the periods of standstill but also other periods of completely erratic light variation.

In general, we find that the latter stars have smaller amplitudes and shorter periods than the U Geminorum variables, although for those stars which show frequent standstills it is often difficult to assign even a mean period. One further problem associated with the majority of the dwarf novae is their faintness, few being brighter than thirteenth magnitude at maximum, and it seems probable that further

149

work on these interesting objects will place more of them in the Z Camelopardalis class than at present.

Spectra of the Dwarf Novae

Just as SS Cygni is the most fully observed of the dwarf novae as far as the light variations are concerned, the same is true of its spectroscopic behaviour, since its magnitude of only 12·1 at an extremely faint minimum makes it a relatively easy object for such investigations. When at maximum the spectrum is virtually continuous, resembling an A1-type star with a surface temperature of about 12,000°K, while at minimum the spectral class is dG5, corresponding to a temperature of only 6,000°K. In view of the relatively short period between maxima, averaging only 54 days, it is difficult, if not impossible, to reconcile this very wide variation in temperature on the basis of a single star.

In 1956, Joy was able to announce that SS Cygni is a spectroscopic binary with an orbital period of only 6 hours 38 minutes, the two components being of spectral types sdBe and dG5. Since it was only two years earlier that Walker has shown Nova DQ Herculis 1934 to be an eclipsing binary with a similar period of 4 hours 39 minutes, it was inevitable that speculation should be aroused that all of the dwarf novae are close binary systems. A detailed spectroscopic survey by Kraft has now proved that several of these stars are binaries with orbital periods of less than 9 hours, providing us with at least circumstantial evidence that all are physical doubles. Here, then, we have a very close link with the novae and, what is possibly more important, in all cases where the spectra of both components can be seen the components are found to be a white dwarf in association with a slightly more massive red companion.

The spectra also provide us with an important piece of additional information. Since the two stars are revolving about their common centre of gravity, it is possible in several cases to measure the Doppler shift of the individual lines due to the approach and recession of each star in its orbit, and from these results we are able to say whether the orbits are circular or elliptical. Curiously, all of the stars belonging to the U Geminorum group have been found to possess circular orbits whereas the two Z Camelopardalis variables for which measurements are available revolve in elliptical orbits. Whether this difference between the two groups will provide us with an explanation of the curious standstills of the latter class is a question that is being actively pursued at the present time. It will be readily appreciated that, since the vast majority of these variables

150

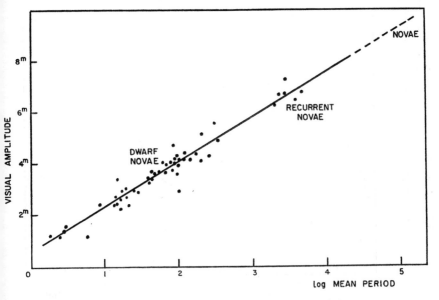

Fig. 31 The period-amplitude curve of the dwarf novae and the recurrent novae.

are extremely faint objects, the problems associated with obtaining satisfactory spectra is a very difficult one.

The Nature of the Dwarf Novae

Until quite recently, very little was known of the nature of these peculiar stars. The major breakthrough was undoubtedly the discovery of their binary nature. The fact that these systems contain a white-dwarf star added some confirmation of the earlier idea that they might be very closely allied to the novae. Here, perhaps, it will be worth while to examine the evidence put forward in support of this view.

The very nature of their light variations inevitably led to their association with the novae, since no other class of star known at the beginning of the century showed changes in brightness in any way similar to those of the dwarf novae. As more data accumulated, attempts were made to find other relationships between the two groups. The first came when a graph of amplitude against mean period was plotted (Fig. 31), which showed that the results for the dwarf novae and those for the recurrent novae fell upon the same straight line. This work was repeated and refined by Kukarkin and

151

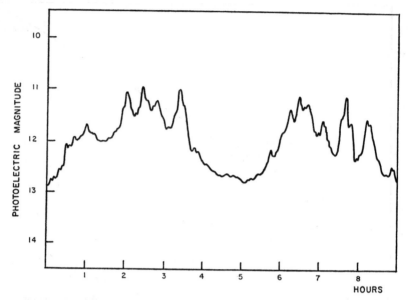

Fig. 32 Photoelectric light curve of AE Aquarii. This peculiar variable, although showing many characteristics of the dwarf novae, has a very complex light variation.

Parenago, who also extrapolated their results to show that such a relationship could also include the ordinary novae on the assumption that their means periods were measured in centuries.

This attractive theory held sway until high-dispersion spectra of the brighter dwarf novae were obtained by means of large telescopes. It was then found that, in contrast to the spectra of the novae, there was no evidence at all for a cloud of rapidly expanding gas surrounding these stars at the time of an outburst and, in addition, the spectra contained none of the 'forbidden' lines that are characteristic of the novae. Reluctantly, the idea that the outbursts of these variables are brought about the same mechanism as is operative in the novae had to be abandoned and some other explanation sought.

What was needed to assist in the interpretation of these odd stars was a working model to which further refinements could be added as more data accumulated. Such a model was provided by a star which, at first sight, does not appear to belong to the dwarf-nova class at all. Although it is not easy to classify this star, AE Aquarii, from its light curve (Fig. 32), the spectroscope shows that it should be included among the U Geminorum variables; furthermore, like

152

them, it is a spectroscopic binary with a period of about 17 hours. It, too, consists of a B-type white-dwarf star and a cool companion of spectral class K5.

Now the mass of the cool star is at least as great as that of the Sun, somewhat higher than normal K-type dwarfs, and its stay on the main sequence will, therefore, be appreciably shorter than for normal stars of this type. The result will be that it is now expanding as it moves away to the right of the main sequence. It is not, however, expanding equally in all directions, as one might expect. The reason for this was elucidated several decades ago, theoretically at first and then confirmed observationally from a study of stars such as those we are now discussing.

If we consider first the simple case of a single star, its gravitational field will extend equally in all directions, this field acting upon all of the atoms that make up the main body of the star, drawing them in toward the centre. Quite obviously, the faster any atom is moving, the farther it will be able to get from the star (ignoring any collisions that may occur with other atoms) and if the velocity is high enough it will, of course, escape altogether. Most of the atoms, however, will be confined within a hypothetical sphere which we may call the zero-velocity or equipotential surface.

Although single stars do not, strictly speaking, possess an equipotential surface, when we extend this argument to the case of close binary systems we find that these do, but now we must take into consideration the interfering gravitational field of the companion star and, instead of a large sphere encompassing both stars, we find that the effect is to distort our hypothetical sphere into a dumb-bell shape composed of two lobes surrounding each component and meeting at a point known as the inner Lagrangian point (Fig. 33).

Suppose now that we apply this idea to a star system such as that of AE Aquarii. We find that the B-type dwarf, which is now in the process of contraction, has pulled all of its mass into a very small volume and consequently does not even fill its lobe of the equipotential surface. The much larger, though only slightly more massive, orange-yellow companion has already filled its lobe and is still evolving, struggling to expand in the process. Clearly, if it is to do so it must force some of its mass through the point of least resistance, this being the inner Lagrangian point L_1, where the reinforcing effect of the blue component's gravitational pull is greatest. The inevitable result is that a stream of gas from the K5 star pours through this point into the unfilled lobe around the white-dwarf star.

We have already seen that this is the case for many of the Algol eclipsing variables. The problem now was to see whether this could

153

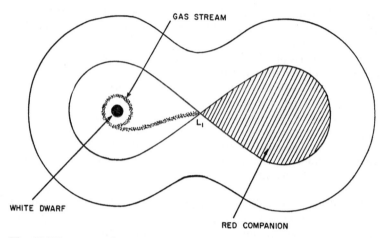

GAS STREAM

L₁

WHITE DWARF

RED COMPANION

Fig. 33 Diagrammatic model of a U Geminorum variable. Owing to the evolution of the secondary component, a stream of gas is being ejected through the inner Lagrangian point L_1.

also be proved for AE Aquarii. Unfortunately such information is not easy to obtain in this case. The star is very faint, and the work is further hampered by the irregular, short-term fluctuations in brightness, which make the spectrum extremely complicated and not easy to decipher. Nevertheless, the existence of such a moving gas stream has been found, and further indirect evidence is provided from a study of the blue star itself. This is found to be much brighter than we would expect for a normal white dwarf and it may well be that some of the gas, drawn into this star from the orange-yellow companion, makes some contribution to its abnormal brightness.

Having this model on which to base their further investigations, astronomers began a study of other U Geminorum stars, all the time working on the assumption, which although without direct observational verification appeared at the time to be correct, that the white-dwarf component, as in the case of the novae, was the seat of the nova-like eruptions.

In 1965, however, Krzeminski was able to undertake a very detailed photoelectric study of U Geminorum in ultra-violet, blue and yellow light at one of its comparatively rare maxima (the mean period of this particular variable is as long as 103 days). Fortunately, this is also one of the few dwarf novae that is known to totally eclipse and he was able to follow the behaviour of the eclipse through the entire maximum. The results turned out to be a little

154

unexpected. During the outburst, the blue star and the rotating gas stream contribute only about 10 per cent of the total light of the system, the rest coming from the cooler component.

It now appears likely that at irregular intervals the convection zone within the atmosphere of the cool star swells rapidly, forcing excess mass through the inner Lagrangian point and at the same time revealing the hotter, lower levels and producing the sharp increase in visible light. All of this evidence that it is the secondary component and not the blue star which is the cause of the outbursts now resolves the difficulty we faced earlier when trying to correlate these variables with the novae. We now see that the eruptions are mainly brought about by exposure of the lower levels and not by a violent ejection of gas, as previously supposed. And since these stars are not generically related to the novae, a further problem is also overcome.

We already know that the majority of the novae are Population I stars belonging to the galactic disc. The dwarf novae, on the other hand, show no preference at all for the regions bordering the Milky Way and appear to be members of an intermediate stellar population, several being known in the neighbourhood of the galactic poles.

The problem of the origin of the dwarf novae has received a good deal of attention in recent years. Comprehensive surveys of their spatial distribution have shown that they resemble one particular kind of star very closely in this respect, and several astronomers now believe that they represent a stage in the evolution of these stars, which are known as the W Ursae Majoris variables after the first of this kind to be discovered. The W Ursae Majoris stars are in reality very close eclipsing binaries which are revolving about each other with their surfaces virtually in contact, the component stars being almost identical and very like the Sun in size, spectral type and mass.

Since one of the components is slightly more massive than the other, it will begin evolving first away from the main sequence, growing larger in the process. This now brings about an interesting situation. Beacuse the two stars are so close to each other, the less massive star begins capturing material from the expanding companion, thereby hastening the other's evolution to the point where it passes rapidly through the red-giant phase and contracts towards a white dwarf. The role of primary and secondary is now reversed; the originally less massive star embarks upon its evolution from the main sequence, expanding in the process and ejecting mass through the inner Lagrangian point. This is exactly the kind of situation we find in the case of the dwarf novae. Clearly, it would be a big point in favour of this attractive theory if we could find stars intermediate between the W Ursae Majoris variables and the dwarf novae. Unfortunately, the changeover from one type of system to the other occurs extremely rapidly, possibly even in the course of a few hundred years, and the chances of finding such a system are very remote.

So far in this discussion we have only briefly mentioned the peculiar standstills of the Z Camelopardalis stars. At present, these remain an enigma. Some last for only a few days whereas others, such as Z Camelopardalis, RX Andromedae, and TZ Persei, have persisted for eighteen months or so with only minor fluctuations. All attempts to find any relationship between the length of a standstill and the brightness of the preceding maximum have proved fruitless, and there is an unpredictability about them that makes it extremely difficult to find any explanation for them whatever. Doubtless future investigations will serve to throw some light upon these mysterious periods of comparative quiescence among these stars.

6. Distance in Astronomy

Astronomy is the science of vast distances, of the ultra-large taking in as it does the furthermost corners of all existence. For long ages, however, it was based solely on what the eye could see and, although this method of observation works efficiently in many cases, for example in the charting of the constellations and the movements of the Sun, Moon and planets across the heavens, it can often lead us into serious error. This is perhaps nowhere more obvious than in the illusion, which persisted from the earliest times to the end of the Middle Ages, that the stars were all at the same distance from us, mere specks of light on an inverted bowl. It is not difficult to understand how this illusion came about. The human eye is adapted to estimating the relative distances of objects only so long as they are comparatively near. When the distance becomes too great, this ability to differentiate between nearby objects and more distant ones breaks down. Where the stars are concerned, a further problem also arises. Prolonged observation with the naked eye reveals no change in their relative positions; all appear to be moving at a constant rate across the sky owing to the axial rotation of the Earth. The belief that the various bodies of the solar system are closer to us than the stars came from their obvious motions against the background of fixed stellar points, but even here it is only comparatively recently that any accurate determinations of distance were made.

We now know that the Moon is a little under a quarter of a million miles away and the Sun some 93 million miles; the nearest star, Proxima Centauri is 4·2 light-years away, the great nebula in Andromeda, 2,200,000 light-years, and the farthest objects visible in the largest instruments, which represent the boundaries of the known universe, are more than 4,800 million light-years distant.

To the layman, such tremendous distances border on the meaningless, and even more difficult to comprehend is the reliance that present-day astronomers place upon their measurements. In reality, the methods used to determine such vast distances are quite straightforward, and in the present chapter we shall be discussing the means whereby astronomers are able to derive not only interplanetary and interstellar distances, but also those to the furthermost galaxies which lie on the very rim of the observable universe. It will be noticed

157

that in the illustrations given above the unit of distance was changed rather abruptly from miles to light-years. This is simply because when we go beyond the confines of the solar system the distances we encounter are so vast that to use miles would result in such large numbers that they would quickly become unmanageable.

Light-Years and Parsecs

A convenient unit of distance, much used by astronomers, is the light-year. This is simply the distance that a ray of light, travelling at more than 186,000 miles every second, will traverse in a year. Measured in miles it amounts to approximately 5,800,000 million – a very long step indeed.

The parsec is a still larger unit of distance than the light-year and has little significance to anyone but the astronomer since it is based upon the measurement of angles. In terms of distance, it is equal to 3·26 light-years. It is more convenient than the light-year since it is readily derived from the parallax; the distance of an object in parsecs is merely $1/p$, where p is the parallax measured in seconds of arc, using the radius of the Earth's orbit (93 million miles) as a baseline.

The Distance of the Moon

Since the Moon is the nearest of all the heavenly bodies, it is only natural that it should have been the first whose distance was measured. The ancient Greeks used the well-known method of triangulation (Fig. 34) to estimate its approximate distance by making simultaneous measurements of its direction from two observatories whose distance apart was known with a fair degree of accuracy. Over the centuries, as more refined equipment became available, this method yielded increasingly accurate results although since the time of Kepler it has been recognized that the orbit of the Moon around the Earth is far from being truly circular. The mean distance between the centres of the Earth and the Moon is 238,857 miles but, since the eccentricity of the orbit is 0·0549, when at its closest (at perigee) it is only 221,463 miles away, whereas at its farthest (at apogee) it is 252,710 miles distant.

The parallax method suffers from certain disadvantages, however, which become more pronounced, as we shall see, when applied to greater distances. As far as the Moon is concerned, the fact that it possesses an appreciable disc means that for the measurements to be accurate it is necessary to choose a particular spot on the lunar surface close to the centre of the disc on which to sight the instruments. In addition, the length of the baseline that may be con-

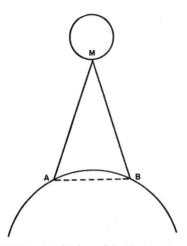

Fig. 34. The triangulation method used by the ancients to determine, by observations from two points A and B on the earth, the distance of the Moon M.

veniently used is limited by the curvature of the Earth itself if the observations are to be made simultaneously, a condition which, in this instance, must be met since the movement of the Moon against the background of the stars is quite rapid: in less than an hour, it travels through an angle equal to its own diameter.

With the development of radar, a more accurate method of determining the distance to the Moon became available to astronomers. Since a radar pulse travels with the velocity of light, it is theoretically only necessary to beam a pulse at the Moon and measure the time taken for it to be reflected back from the lunar surface in order to determine its distance. The United States Army Signals Corps made radar contact with the Moon for the first time on January 10, 1946 by utilizing the directive effect of an array of antennas. Each pulse, sent at 5-second intervals, returned after a delay of 2·56 seconds. With modern techniques of time measurement (for it is the accuracy of the timing that is important here), it is now possible to determine the distance of the Moon with a high degree of accuracy.

Even the radar method has now been superseded, however, by what will almost certainly prove to be the ultimate in distance measurement as far as the Moon is concerned. One of the tasks undertaken by the astronauts during the Apollo 11 Moon landing was to set up a highly efficient reflector on the lunar surface in the Mare Tranquillitatis. By sending a laser pulse to the Moon and

159

bouncing it back from this reflector, it is now possible to measure the distance to the lunar surface with an accuracy of about five feet.

Distances within the Solar System

The distances of most of the planets are too great for the parallax method to be used, as it is for the Moon. Only Venus and Mars are sufficiently close for approximate values to be obtained. The longest baseline we can use for the determination of planetary distances is always something less than 7,900 miles – the diameter of the Earth – and unless the object is very close it will not have an accurately measurable parallax. However, it is not necessary to measure the distance to each individual planet. To see why this is so, we must mention here Kepler's third law of planetary motion which states that the square of a planet's period of revolution about the Sun is proportional to the cube of the major axis of its orbit, this being true of every planet irrespective of the dimensions of its orbit.

This law, together with the fact that every body within the solar system, whether it is a planet or an asteroid (one of the tiny bodies which revolve mostly in orbits between Mars and Jupiter), obeys the law of gravity, means that we have only to determine the distance to any one body to be able to determine the scale of the solar system and calculate the distances to all of the others. We cannot use the Earth–Moon distance since, although the Moon moves around the Sun with the Earth, the dimensions of its orbit that we measure are those around the Earth and not the Sun. Fortunately, Venus and Mars are not the only candidates. Several of the asteroids, moving in elongated ellipses, sometimes approach much closer to the Earth. In 1931 one of these minor planets, Eros, came within 16,200,000 miles of the Earth and astronomers prepared for this event by making accurate star measurements of the region where Eros would be during its close passage to the Earth. These provided an excellent background against which to make the measurements. The results obtained have been only slightly modified by later determinations made on other asteroids which approach even closer to the Earth than Eros.

The Dimensions of the Galaxy

When we go beyond the boundaries of the solar system, the measurement of astronomical distances becomes more difficult and complex, yet it is still based upon well-known and well-tried principles. Until the beginning of the nineteenth century it was widely believed that the stars were so remote that it was quite impossible to

160

determine their distances. All that could be said was that they were far more distant than the planets since they showed no apparent motion among themselves. Attempts were made towards the end of the eighteenth century to use the method of trigonometrical parallax which had yielded good results for certain planetary distances, to determine the distance to certain of the stars, but with a marked lack of success.

At this time, too, it was considered that brightness was a criterion of nearness, this idea being based upon the erroneous belief that all of the stars are of about the same intrinsic luminosity. We now know that this is far from being the case, and many of the brightest stars are so remote that the parallax method fails utterly as a means of measuring their distance even with present-day instruments and techniques. In addition, at the time, it was thought that the longest baseline which could be used was the diameter of the Earth, a little over 7,900 miles.

It was not until two important facts were recognized that it proved possible to measure stellar distances by the parallax method. The first of these was that an even longer baseline than the diameter of the Earth exists from which we can make our measurements, namely the diameter of the Earth's orbit about the Sun, a distance of 186 million miles. In order to utilize this baseline, all that is necessary is to make measurements on the star at six-monthly intervals.

The second was that the stars themselves are moving through space in random directions and with widely different velocities. Our knowledge of these motions is comparatively recent simply because they are not apparent to the naked eye and it is necessary to make a series of accurate observations over quite a long period before they become noticeable. Although the very small displacements of certain stars as seen against the more remote background – the 'proper motions' of the stars – were discovered by Halley as long ago as 1718, it was over a century before this discovery was used for the determination of stellar distances.

In 1838, Bessel chose the 5·5-magnitude star 61 Cygni as a likely candidate, basing his choice upon the assumption that its large proper motion across the heavens was indicative of its comparative closeness to the Sun. Using the trigonometric parallax method, he was able to establish its distance as almost eleven light-years. Since this first breakthrough, this method has been successfully applied to several nearby stars, the nearest being Proxima Centauri, visible only from the Southern Hemisphere. This red-dwarf star lies at a distance of only 4·2 light-years and is thought to be a companion of α Centauri, which is at approximately the same distance (4·3 light-years), the two revolving about each other in something like a million years.

With increasing distance, measurement of the very small angles involved becomes much more difficult and the parallax method is reliable only up to distances of about 330 light-years. Stars more remote than this are beyond the range of such measurement with present instruments and other methods have to be used.

One such method, that of dynamic parallax, is applicable only to binary stars whose individual components can be separated in the telescope. Here, if we assume that the masses of the component stars are similar to that of the Sun (an assumption that is approximately true in the majority of cases), we can use the law of universal gravitation to calculate the linear diameter of the orbit. Once we have done this, we need only compare it with the observed angular diameter to immediately obtain the distance. Fortunately there are quite a large number of such visual binaries to which this method may be applied and a great deal of valuable statistical information has been obtained from their study.

A further method which may be used for distances that are too great for direct measurement by trigonometric parallax depends upon the proper motions of the stars themselves combined with that of the Sun. Everyone is familiar with the effect of motion on distance as seen from a railway carriage. Nearby objects move past very rapidly compared with more distant objects on the skyline. When we apply this method to the determination of astronomical distances, however, we are faced with the additional difficulty that the objects whose distances we are trying to measure are themselves moving with widely different velocities. We must therefore determine the proper motions of a large number of stars in order that the differences in their velocities may be averaged out. Once this has been done and the absolute velocity of the Sun itself thereby established, the distances of groups of stars may be readily found from the mean annual proper motion of each group. This method is known as that of hypothetical parallax and has recently come into prominence in the measurement of stellar distances.

When we mentioned the use of the Cepheid variables as distance indicators in Chapter 2, a brief idea of the difficulties associated with this method was given, not least being the problem of finding the distance to any one Cepheid in order to establish the zero point for the period-luminosity relation. Then we saw how spectroscopic methods might be used for such a determination, although in this particular case it was not possible. The method of spectroscopic parallax is, however, one of the most fruitful of all. Basically, it

162

involves determining the absolute magnitude of a star from its spectroscopic characteristics, especially the intensities of the absorption lines in the spectrum and where the maximum radiation lies in the continuum, the latter providing us with a fairly accurate guide to the surface temperature. Once the star's absolute magnitude is known the distance may be easily found by comparison with its apparent magnitude.

The spectroscopic parallax method has one advantage over all of the others we have mentioned so far. With the exception of the nearer stars, for which the trigonometric parallax method gives reliable results, all other methods yield increasingly uncertain figures as the distance increases. The spectroscopic method, however, does not change in accuracy with distance, any uncertainty remaining virtually constant, and it is limited only by the ability to recognize the features of the spectrum.

RR Lyrae and Cepheid Variables as Distance Indicators

In Chapter 2 we introduced the idea of utilizing the RR Lyrae variables and the Cepheids as distance indicators and it is now opportune to discuss this method in more detail. The RR Lyrae stars have very distinctive light variations and since they are also about a hundred times brighter than the Sun they are readily picked out at great distance. Since they occur in large numbers in the globular clusters, these stars have also been termed the cluster variables. When we examine their light curves we find that the majority have periods lying between 0·3 and 1·3 days and, since they all have approximately the same intrinsic brightness, we can clearly use them as a standard means of deriving distances. The early measures were all based upon an absolute magnitude of 0·0 for the RR Lyrae variables, but more recent determinations by Miss Fringant, using a special spectroscopic technique devised by Chalonge, showed that these stars are somewhat fainter intrinsically that at first thought and their absolute visual magnitudes are nearer +0·8. This redetermination of their luminosities has resulted in a general reduction of the distances earlier deduced for the dimensions of the Galaxy.

Since we are immersed within the Galaxy, it is not easy for us to visualize either the structure or the size of this great collection of stars. The early ideas of Sir William Herschel, who showed that it possesses a lenticular shape, have been modified in certain respects. We now know that this picture is basically true as far as the general system of stars goes. The overall diameter of the lens-shaped system is of the order of 100,000 light-years, with the Sun near the periphery, about 25,000 light-years from the centre, within one of the spiral arms. The central regions also form a very pronounced bulge about 8,000 light-years across.

What was not realized by Herschel, however, was that there is also a very extensive halo of globular clusters around the Galaxy, whose distances can be measured by means of the RR Lyrae stars they contain. At a distance of about 120,000 light-years from the galactic centre there are also a fair number of star systems which may be regarded as satellites of the Galaxy, since they almost certainly circle it in a period amounting to some 2,000 million years. These stellar systems are somewhat larger than the globular clusters, consisting of several million stars as compared with about 100,000 in a typical globular cluster.

A recent survey carried out at Mount Palomar has revealed the presence of several of these satellite systems. Significantly, the Andromeda nebula, illustrated in Plate X, has two very noticeable satellites, which are fairly typical elliptical galaxies and sufficiently large to have received individual catalogue numbers – NGC 205 and NGC 221. These are, of course, much larger than any of the satellite systems of our own Galaxy. Recent calculations by Schwartzschild have shown that the total mass of these two systems of the Andromeda nebula is about 5 per cent that of the main spiral, whereas the total mass of all the satellite systems of our own Galaxy amounts to only about 0·1 per cent of the main system. The Andromeda spiral also possesses a halo similar to our own but from a distance of 2,200,000 light-years – this is too faint to be readily photographed.

The RR Lyrae variables have also been used to determine distances beyond our own system but not as far out as the Andromeda spiral. There are several galaxies nearer to us than M 31 in Andromeda although they are mostly faint objects. The distances of the Magellanic Clouds have been measured by means of these stars, particularly by Thackeray and Wesselink, who found a value of about 180,000 light-years. We shall see when discussing radio astronomical techniques that the two Magellanic Clouds form a connected system, moving in an orbit around a common centre of gravity and the view that they are also moving about the Galaxy and are therefore true satellites is generally accepted.

The Cepheid variables, possess special characteristics which enable us to use them as distance indicators. Indeed, they may be used for even greater distances than the RR Lyrae stars, since they may be up to 300 times brighter than the stars of this class. However, the problem here is a little more complicated since they do not form anything like a uniform group.

We have already seen that as long ago as 1912 Miss Leavitt

164

discovered the period-luminosity law which applies to the Cepheid variables, her researches being carried out on the large number of these stars found in the large Magellanic Cloud. Unlike the RR Lyrae variables, these stars occupy a zone on the Hertzsprung–Russell diagram.

At first the distinction between the RR Lyrae variables and the Cepheids was not recognized and it was considered that both classes of star obeyed the same period-luminosity law. On this basis, the RR Lyrae stars and the Cepheids were used indiscriminately for determining distances, first to the Magellanic Clouds and then out to M 31 in Andromeda. It was at this point that the first anomalies were encountered since from this initial survey the distance to the Andromeda Galaxy was calculated to be only about a million light-years.

When the 200-inch reflector came into operation, it was found that whereas the photographic limit of this instrument is about magnitude 24, there were no RR Lyrae stars to be seen in this galaxy. If the distance of only a million light-years was correct, then such stars should have been visible in large numbers as objects of photographic magnitude 22·4, well within the range of the Mount Palomar reflector. Either there were no RR Lyrae stars within this huge stellar system (and since in every other respect it closely resembles our own this seemed highly unlikely) or M 31 was much farther away than had been calculated by means of the Cepheid scale as calibrated from the RR Lyrae stars.

To get over this difficulty, Baade used the long-period variables which had already been identified in the Andromeda spiral by means of their late-type spectra and characteristic light fluctuations. Here it was discovered that they had a mean photographic magnitude of 22·4 and, since astronomers had previously found from similar stars in our own Galaxy that their mean absolute magnitude is around − 1·5, Baade was forced to conclude that the Andromeda nebula is at least twice as distant as had been previously calculated. Most other intergalactic distances also had to be doubled, the Magellanic Clouds being 180,000 and the Andromeda Spiral 2,200,000 light-years distant respectively.

More recent work, using the novae and bright blue stars in the nearer galaxies, has amply confirmed these results. One further anomaly which had also troubled astronomers was cleared up at the same time. The earlier-adopted measurements had indicated that our own Galaxy was a veritable giant among the galaxies of the Local Group, far larger than even the biggest among them. When the dimensions of these galaxies were worked out using the fact that not only are the Cepheids quite distinct from the RR Lyrae stars but also do not fall on a line on the Hertzsprung–Russell diagram but within a zone which is roughly 1·5 magnitudes wide, the modified values were much more reasonable. Not only had Baade effectively doubled the size of the known universe at that time, he had also reduced the overall dimensions of the galaxies to their proper proportions.

The Cepheids can be used as distance indicators up to about 10 million light-years. The somewhat brighter long-period variables and novae virtually treble this distance before they become too faint to be identified. Within this vast volume of space lie about a thousand galaxies of various types whose distances can theoretically be measured by means of single stars of the kind just discussed, although at the present time only a small fraction of these distances have actually been measured.

Even 30 million light-years, however, is a very small step compared with the distances in which astronomers are interested. The great clusters of galaxies, as we shall see later, stretch away to the very edge of the observable universe, many thousands of times more remote than this. In order to extend our measurements we clearly need something far brighter than a single star. One possibility that immediately suggests itself is to use the integrated light of a globular cluster which, as we have seen, contains up to 100,000 stars all compacted together into a relatively small volume. A programme of distance measurements based upon the globular clusters is in progress at present but there are several difficulties which must be overcome before reasonably accurate results can be obtained. Not least of these is the problem of calibrating the intrinsic brightnesses of these clusters for, unfortunately, they cover quite a wide range of luminosities.

At very great distances even the globular clusters are too faint to be seen and we must then fall back upon the galaxies themselves. Astronomers have been able to use the various types – irregular, spiral and elliptical – to measure these extremely large distances, basing their figures upon the average brightness of each type. Like the globular clusters, the galaxies cover a range of sizes and luminosities, but when we consider a fairly large cluster of them, presumably containing examples of every size and brightness, these variations tend to cancel each other out, yielding a mean figure that is accurate to within 15 or 20 per cent of the true distance.

Additional confirmation of these figures comes from a different source, namely the red shift observed in the spectra of the galaxies. When we come to discuss the expansion of the universe, we shall see that this Doppler shift of the absorption lines in the spectra of a galaxy (from which we obtain its velocity of recession) is directly proportional to its distance and provides us with a check on the values obtained by other methods. It will be noticed that here we have mentioned only velocities of recession and not of approach. Only the various galaxies which belong to the Local Group show a blue shift of their spectral lines; all others, without exception, are moving away from us with high velocities.

We can now give a general summing up of the various methods used by astronomers to measure the distances involved in the whole field of astronomy from the extremely small dimensions of the solar system out to the furthermost points of the universe as we know it. Beginning with reliable methods that yield extremely accurate results, we can move outward, step by step, progressing from interplanetary distances, through the huge stellar system that is our Galaxy, and thence out into the region of the galaxies. Inevitably, the farther out we go, the greater is the uncertainty in our results; nevertheless, each step we take is based logically and firmly upon the one that precedes it.

In the space of half a century, our conception of the universe has grown from that of an isolated system – the Galaxy – to a seemingly boundless universe containing as many galaxies as there are stars within the Milky Way. With this tremendous rolling back of the astronomical horizons have come other problems, many of which still await solution and lead us down strange paths which astronomers are only now beginning to explore. These, however, we must leave until a later chapter.

7. Beyond the Visible

The light by which we see the Sun, Moon, planets, stars and even the furthermost galaxies all comes through a very narrow range in the electromagnetic spectrum, extending from approximately 3,900 Å in the violet to 7,400 Å in the red (Fig. 35). These limits depend upon the sensitivity of the human eye, which varies appreciably from one individual to another. As long ago as 1800, however, Sir William Herschel used a thermometer with a blackened bulb to show that there exist other, invisible, radiations in the spectrum of the Sun, lying beyond the red end of the visible range, which produce a heating effect. These are known as the infra-red radiations, and by means of special photographic plates it is possible to trace them out to about 12,000 Å. Other means enable us to detect them to the limit of the solar spectrum as far as 50,000 Å. Radiation of even longer wavelengths lies beyond the infra-red, these being known as the Hertzian or radio waves.

In 1801 Ritter proved that the visible spectrum extends in a similar manner beyond the violet end. These ultra-violet radiations, like the blue and violet portions of the visible spectrum, are capable of bringing about certain chemical decompositions. For example, they will affect ordinary photographic emulsions, which may be used to record them out to about 2,000 Å, while special apparatus can trace them down to 140 Å.

The electromagnetic spectrum does not, of course, stop here. X-rays, which have proved so useful in medical science, cover the region from 25 Å to 0·1 Å, merging imperceptibly into the shorter γ-rays, which go down to below 0·05 Å. Finally, the electromagnetic portion of cosmic radiation, which apparently permeates the whole of space, has wavelengths as short as 0·0004 Å.

Many of these radiations have a destructive effect upon living beings. Ultra-violet light of high intensity is lethal to animal life and hard X-rays have a sterilizing effect upon both plant and animal life. Fortunately for our continued existence on Earth, only a small proportion of the radiation from the Sun penetrates the atmosphere. Both oxygen and a relatively thin layer of ozone lying at the outer fringes of the atmosphere absorb all radiation with wavelengths shorter than about 2,900 Å in the ultra-violet, except the highly

168

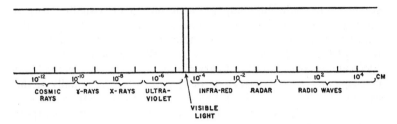

Fig. 35 The electromagnetic spectrum.

penetrating cosmic rays, and less than 1 per cent of the solar radiation in the infra-red having a wavelength greater than 40,000 Å is transmitted by atmospheric water vapour. A similar situation, of course, holds for the radiation reaching the Earth from the stars. As a result, if we wish to examine the Sun or stars at these very short or long wavelengths, we must have some means of viewing them from outside the atmosphere. The radio waves, on which the science of radio-astronomy is based, are the only exception.

Until a few years ago, the only way we could observe the heavens outside the visible range was to send up quite small instruments in balloons which carried them into the outer reaches of the atmosphere, sufficiently high to be above most of the absorbing layers. Now, however, it is possible to put artificial satellites into orbit around the Earth carrying much larger and more complex instruments and remaining in orbit for far longer periods, enabling comprehensive surveys at these various wavelengths to be carried out. At some time in the future, small observatories will be erected on the lunar surface which will initially be operated by signals sent from Earth and eventually be manned by scientists.

The study of the stars at wavelengths outside the normal visual range, either by means of special telescopes and ancillary equipment on Earth or by instruments carried by orbiting satellites, has received much attention in the last few years and has been productive of a greal deal of extremely valuable information. It is proper, therefore, that the present chapter should be devoted to the results that have so far been obtained.

The Infra-Red Stars

For a long time now, astronomers have realized that the heavens would look vastly different to their present appearance as seen by the naked eye if it were possible to view them in infra-red light. Perhaps the first suggestion of this was obtained when the early photographic

169

surveys were made and compared with those made visually. Normal photographic emulsions have quite a different sensitivity to the various colours from that of the human eye. The maximum sensitivity of the eye lies somewhere in the yellow region of the visible spectrum whereas that of ordinary photographic plates is in the blue; consequently a red star such as Betelgeuse, which is visually of the first magnitude, is scarcely visible on a photograph. In the same way a blue star, Rigel for example, has a brighter photographic magnitude than a visual one.

Most, if not all, of this difference is due to the surface temperature of the star in question. The sun has a surface temperature of 5,700°K and its maximum energy lies around 5,000 Å in the yellow region of the spectrum. Other stars are known, mostly long-period and irregular variables, that have temperatures between 2,000°K and 3,000°K, most of their radiation coming through in the red end of the visible spectrum; those which are cooler still (only a few of which are just visible to the eye) emit the greater proportion of their energy in the infra-red. The wavelength of the maximum radiation given out by a hot body such as a star is inversely proportional to the surface temperature, as shown in Fig. 36, from which it will be seen that from those 'stars' whose surface temperatures are below about 1,000°K, no visible light at all is emitted.

Even before these infra-red sources were found, a simple argument suggested that they did indeed exist. Suppose that we examine the relative numbers of stars of different types in any particular volume of space. If we ignore such regions as the Orion nebula, which is known to be teeming with young, hot stars recently born out of the gas and dust clouds, we find that most of the stellar inhabitants in space are cool stars, either red dwards or red giants with low surface temperatures. This would indicate that there must be large numbers of stars which cannot be seen, either visually or during ordinary photographic surveys, simply because their surface temperatures are so low that the amount of visible light they emit is negligible, and the only means of detecting their presence is to search for them in the infra-red region of the spectrum.

Obviously it is more difficult to detect such infra-red sources than it is to locate visible ones. However, there are two methods by which this may be done. The first, employed by Hetzler some thirty years ago, used special photographic plates sensitive to infra-red radiation. The preliminary survey carried out by Hetzler with the 40-inch refractor at Yerkes Observatory was successful in locating several stars that have surface temperatures as low as 1,000 to 1,500°K.

A more elegant method has recently been devised by Neugebauer

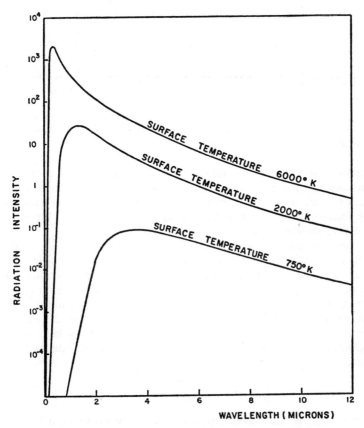

Fig. 36 Diagram showing how the peak of the radiation from stars depends upon their surface temperature. Below about 1,000°K no visible light is emitted.

and Leighton who have constructed an infra-red telescope having a diameter of 62 inches and a short focal length of only 64 inches. Basically the instrument is a reflecting telescope, but in place of an aluminized glass mirror the parabolic surface is composed of an epoxy resin on an aluminium disc, the resin being finally coated with aluminium in the ordinary way. At the focus of the instrument is an infra-red detector which is cooled to a low temperature by liquid nitrogen.

At this point it is necessary to examine the particular wavelengths that may be used in infra-red work using instruments on the Earth's

171

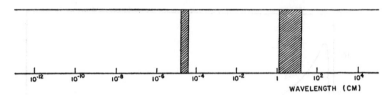

Fig. 37 At the surface of the Earth only two narrow regions of the electro-magnetic spectrum are observable—the visible region and a somewhat broader portion of the radio spectrum.

surface. As mentioned earlier, the water vapour present in the atmosphere absorbs much of the longer wavelengths beyond the red end of the visible spectrum and there are only a few very narrow regions in the spectrum where it is sufficiently transparent for useful observations to be made, as may be seen from Fig. 37.

Another problem encountered by astronomers working in the infra-red region is that of radiation from surrounding objects. We ourselves are not particularly aware of its existence because our eyes are completely insensitive to it, although we do, of course, feel the heat rays in the near infra-red from warm or hot objects. We may readily calculate where in the spectrum the peak of the emitted energy from a body will occur by means of Wien's law, which relates the maximum of the radiated energy to the absolute tempera-ture by the equation:

$$\lambda_{max} = \frac{3,000}{T} \tag{1}$$

where λ_{max} is the wavelength measured in microns (1 micron is 0·001 millimetre) at which the maximum spectral distribution of the radiated energy occurs and T is the absolute temperature (degrees Kelvin). From this we can readily see that in the case of the Sun, with a surface temperature of 5,700°K, the maximum corresponds to about 0·5 microns (5,000 Å). Similarly, for a body at normal room temperature, about 300°K, the peak of the emitted energy comes at 10 microns, in the infra-red. The situation has been likened to that of an optical astronomer forced to work in the open in broad daylight.

This difficulty has been overcome by Neugebauer and Leighton in a very ingenious manner by rocking the mirror at a constant rate, thereby moving the image of the extra-terrestrial source on and off the infra-red detector, thus producing in it an alternating current. The background radiation, on the other hand, covers most of the focal plane of the telescope and therefore provides a nearly constant current. By the use of suitable amplifying circuits it is possible to smooth out the latter and enhance the former.

The wavelength range chosen by Neugebauer and Leighton was the narrow region of the infra-red between 2·0 and 2·4 microns and their sky survey has already resulted in the discovery of some 20,000 infra-red sources, of which 5,000 have been catalogued, being approximately as bright in the infra-red as are the 6,000 or so naked-eye stars in the visible region.

How does the infra-red sky compare with the familiar constellations as seen by the naked eye? Since only a relatively small number of the visible stars radiate strongly in the infra-red (those which are either very cool objects or distant stars whose light has been reddened by passage through interstellar dust), the infra-red sky is very different to that which we know and few, if any, of the well-known constellations are identifiable.

Distribution of Visible and Infra-Red Stars

From the survey made by Neugebauer and Leighton, several interesting results have emerged, particularly as regards both the structure and the dimensions of our Galaxy. When the heavens are examined with the naked eye on a clear, moonless night it is immediately obvious that the brighter stars are distributed more or less randomly in the sky. These are merely the few stars which lie within a sphere having a diameter of about 3,000 light-years with the Sun at the centre. The faintest stars we can see are concentrated along the plane of the Galaxy, forming the Milky Way with its centre lying in the constellation Sagittarius. This is, of course, what we would expect to find. The bright stars are all relatively close to us and we find approximately the same number in the line of sight no matter in which direction we look. The fainter ones, however, which are in general the more distant ones, thin out appreciably in the direction of the galactic anticentre but show a marked increase in numbers towards the centre. All of this has been known for three centuries, ever since the telescope resolved the Milky Way into innumerable stellar points and provided the first true picture of the shape and size of the Galaxy.

A similar distribution has been found for the infra-red stars. Once again, the brightest of these objects show a randomness which parallels that of the visible stars and for the same reason, namely that when we consider only a small volume of space near the Sun the Galaxy is essentially uniform. The distribution of the faint infra-red sources shows a much more pronounced preference for the plane

173

of the Galaxy, especially in the direction of the centre, than is found for the visible stars. This does not necessarily imply any fundamental difference in these two groups of stars. It is much more likely that the reason lies in the ability of the infra-red astronomer to see much further than his optical colleague.

The space between the stars is not empty but contains both gas and dust. Recent investigations have demonstrated that the density of this material varies widely from one region of the Galaxy to another. As far as the Milky Way is concerned, the absorption of starlight by this interstellar gas and dust amounts to at least one magnitude for every kiloparsec in the direction of the centre of the Galaxy and about half a magnitude per kiloparsec towards the anti-centre. Such a dimming affects infra-red stars but by no means to the same extent. If, for example, we consider those stars which lie at the very centre of the Galaxy within the dense dust clouds, we find that the visible light is diminished by a factor of about 10,000 million whereas the infra-red radiation centred at 2·2 microns, emitted by red giant stars in this region, is dimmed only by a factor of 10.

Quite clearly, by means of infra-red telescopes in the 2·0–2·4 micron window of the spectrum, it is possible to examine such stars at far greater distances than can be done in visible light. There still remains, of course, the difficulty of differentiating between the redness that is due to the low surface temperatures of nearby stars and the well-known interstellar reddening of very distant objects. Nevertheless, the problem is not quite as complex as in the case of visual observations.

All of this work, which is still being energetically pursued, has added greatly to our knowledge of the size and structure of the Galaxy. Measurements have shown that the diameter of the galactic nucleus is in the region of 4,000 parsecs (13,000 light-years), which is in quite good agreement with values obtained by other methods. This ability to penetrate farther through the obscuring dust clouds that lie toward the centre of the Galaxy has also provided direct proof of the thinning out in numbers of stars as we proceed towards the periphery of our stellar system.

In one respect, however, these surveys have failed to give us an answer to the problem that has faced astronomers for several decades and that it was hoped this infra-red technique would resolve. The pioneering work of Kapteyn and Boss indicated that the stars of the galactic plane are streaming in two preferred directions, toward Orion and Sagittarius. The explanation of this was given in 1927 by Lindblad, who showed that the Galaxy as a whole is rotating; later,

174

more refined calculations have yielded a rotational period of approximately 200 million years for the Sun, which lies at some 33,000 light-years from the centre. Knowing this, we may work out the total mass of the Galaxy within fairly close limits. Unfortunately, when we determine the density of matter in the Galaxy as assessed by totalling the known amount of interstellar gas and dust, and the contribution from the stars, it was found that figure obtained is less than the density of gravitating matter near the Sun based on studies of stellar motions at right angles to the plane of the Milky Way. This suggested that almost half of all matter in the solar neighbourhood remained undetected and this became known as the missing matter.

Where, then, was all of the missing mass? Unless some very serious underestimates had been made in the early calculations, the most probable answer seemed that it must lie in dark stars that cannot be detected either visually or photographically. By the term 'dark stars' we do not mean those which have necessarily reached the end of their careers and are now totally dead, inert globes of matter, emitting no radiation whatsoever. From what we know of the age of the Galaxy and the time spent by a typical star from birth to eventual death, it is doubtful that the number of such stars can be very great. It seemed much more likely that these stars are those which are so cool that little, if any, of their light comes through in the visible region of the spectrum. This, of course, is where it was anticipated that infra-red surveys of the Galaxy might overcome this curious anomaly, providing us with some idea of the total number of such cool bodies and therefore a measure of their combined mass. Faint infra-red stars have been found, the majority of them lying among the obscuring dust clouds near the galactic centre, but their total mass forms only a very small fraction of the combined mass of the Galaxy itself and the problem was not solved until Woolley and his colleagues measured the velocities of A-type stars fainter than those previously determined. This showed that there are two kinds of such stars, one having a greater velocity of motion through the Galaxy than the other. Calculations based upon the amended density in the solar neighbourhood are in accord with that derived from all known galactic matter and consequently there is no missing matter.

Protostars and Planetary Systems

The existence of patches of dark nebulosity extending through large volumes of space has been recognized for several decades. The Horsehead nebula in Orion (see Plate XIII) is an excellent example

175

of such clouds of obscuring matter, which are widely distributed throughout the Galaxy. In addition to these large nebulosities, however, certain much smaller and essentially spherical globules have recently been discovered that are only one light-year or so in diameter. The density of the material in these small nebulosities appears to be inversely proportional to their size and it is noticeable that those found in the rich stellar regions of the Milky Way are somewhat smaller than others in the opposite regions of the sky. Since they emit no visible light and are so small, they are best seen when projected against the background of the diffuse gaseous nebulae, such as Messier 8 in Sagittarius (see Plate XIV).

Many astronomers are now of the opinion that these globules are clouds of gas and dust that are condensing under the combined influence of gravity and the intense pressure of radiation from the surrounding stars – in fact, that they are the first stage in star formation. Whatever their nature, their temperature must be very low, probably only of the order of 400 to 500°K, only a little above that of boiling water and as a result virtually all of their radiation will be in the infra-red. They are thus ideal objects for examination in infra-red light and several surveys have been made at various wavelengths in a search for such protostars.

Plate XV shows one such source in Cygnus, close to the second-magnitude star γ Cygni which, although visible on red-sensitive plates, is completely absent when the region is photographed in blue light. As we move further into the infra-red, however, the brightness of this object increases. At a wavelength of 2·0 microns it rivals Vega, the brightest of all the northern circumpolar stars, and at 20 microns it is brighter than any other object apart from the Sun and η Carinae. At the moment it is impossible to reach any firm conclusion as to the nature of this object, which is peculiar in the extreme. The temperature of this infra-red source appears to be around 1,000°K which would place it among the cooler of the long-period or semi-regular stars, most of which have surface temperatures in the range 1,500 to 2,500°K. However, accurate measurements have shown that this object does not exhibit any variability in its output of radiation so clearly it does not belong to the class of variable stars.

Penston has argued that during the early period of star formation the condensing protostar will be surrounded by a shell of gas and dust, which will inevitably be at a lower temperature than that of the forming star itself. Such a picture would fit the Cygnus source although it is probably significant that no other young stars have been located in its vicinity as would be expected, since modern ideas

176

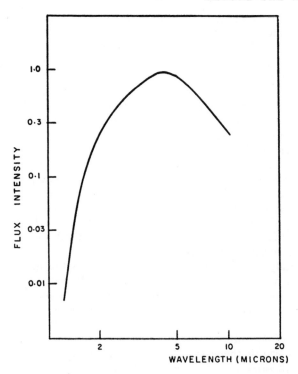

Fig. 38 The infra-red radiation curve of the point source within the Orion nebula. Very little radiation indeed comes through in the visible spectrum.

of star formation suggest that they are formed in groups and not singly.

On the other hand it may be a very bright, hot star of the super-giant class situated at a great distance from us, the light being abnormally reddened by interstellar dust. Even this idea faces certain criticisms which are not readily explained. The main objection is that so far no other supergiant stars like this have ever been observed.

Two other objects worthy of detailed mention are a recently discovered point source within the Orion nebula and R Monocerotis, a peculiar variable star that has been known for several decades. The first-mentioned of these infra-red sources was located by Becklin using the 60-inch reflector at Mount Wilson Observatory. No visible light whatever is emitted by this object and it is totally invisible on photographs which show stars as faint as magnitude 21.

M   177

The peak of the radiation curve occurs at 4 microns (Fig. 38), corresponding to a black-body temperature of only 650°K.

This probable protostar has been the subject of investigation by several groups with very interesting results. In 1968, the region was examined at a wavelength of 20 microns by Low and Kleinmann, who found no trace of the object but did discover, very close to its position but apparently separated from it, a much larger and brighter source. At 22 microns, this region is almost as bright as the full moon is visually!

Further investigations made more recently have suggested that the temperature of this extended source is as low as 150°K – 120C deg. below the freezing point of water – and that the point source mentioned earlier is emitting radio waves characteristic of the hydroxyl radical. The whole question of radio emissions from various atoms within the Galaxy will be discussed in detail in a later chapter. Here we may simply state that the hydroxyl radical is usually found closely associated with the presence of water.

Naturally, the question now arises: what is the nature of the infra-red point source in the Orion nebula? It may, of course, be simply a star that is deeply embedded within the nebula and whose light has been reddened by the gas lying between it and ourselves, the visible light having been both scattered and absorbed before it reaches us. The other possibility is that this source is an extremely cool protostar in the process of condensing out of the dust cloud. Of the two theories, the latter appears the more probable. If the object is merely a normal star deep within the nebula, then we may make certain calculations to estimate the amount of dust that would be necessary to reduce its light by the observed amount. When we do this we find that it would require a thickness of dust and gas much greater than the known diameter of the entire Orion nebula. If, however, the source is a star in the process of formation, then we may make some reasonable assumptions regarding its probable mass and diameter which indicate that we ought to be able to detect changes in both temperature and size in the course of a few centuries which, astronomically speaking, is a very short time indeed.

The remarkable variable star R Monocerotis has been observed for over half a century and a great deal of information regarding its visual light variations has been accumulated. Visually, the star varies in quite an unpredictable and irregular manner and is generally classed as an RW Aurigae variable: these are closely akin to the T Tauri stars which, as we have already seen, are very young objects associated with small patches of dark nebulosity. The spectral type is approximately dG5 and, like the majority of these stars, this one

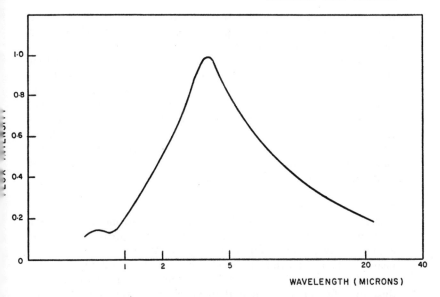

Fig. 39 Diagram showing the peak of the infra red radiation emitted by the peculiar long-period variable star R Monocerotis.

is also associated with a small nebulosity, the light of which is itself variable. At times, the variations in brightness of the nebula keep in step with those of the star itself but at others the light changes appear to be completely independent and it has been suggested that very small changes in the variable are accompanied by much larger variations in the infra-red and ultra-violet radiation emitted by the star which have a pronounced effect upon the luminosity of the associated nebulosity.

R Monocerotis has a visual range of magnitude 10·0 to 14·0 and nothing out of the ordinary was suspected regarding this star until 1966 when Mendoza, working at the Tonanzintla Observatory, discovered a second, and much larger, peak in the spectrum of the variable, centred at 3·8 microns (Fig. 39).

This discovery has led to a dramatic reappraisal of the nature of R Monocerotis. Originally, on the basis of the optical spectrum alone, the surface temperature was estimated at about 5,500°K, very like that of the Sun. Now that it has become evident that the greater proportion of its radiation comes in the near infra-red, this figure has been drastically reduced and it is now believed to be in the region of only 750°K. We must bear in mind, however, the very peculiar

179

nature of this variable and it seems likely, as Low and Smith have pointed out, that here we have a very young star which is surrounded by a dust cloud that absorbs the shorter wavelengths and re-radiates the energy in the infra-red. There is some evidence to suggest that this dust cloud is not in the form of a spherical shell but rather a disc lying in the plane of the star's equator. From spectroscopic evidence, there is no indication whatever that the star is accreting mass from this disc; rather the opposite appears to be the case and it is probable that here we have a planetary system in the making. This is quite a feasible proposition since the T Tauri variables, as a class, are among the youngest of all the stars, almost certainly still in the process of evolving onto the main sequence.

The Centre of the Galaxy

Before leaving this discussion of infra-red astronomy, we must examine one particular field in which it has proved to be an extremely powerful tool for elucidating the structure of our own and nearby galaxies, particular the closely-packed central regions. We have already seen that infra-red radiation is absorbed by gas and dust to a far smaller extent than visible light and it is therefore possible to use this property to observe the fine structure of the galactic nucleus.

Photographs of several galaxies have firmly established the presence of highly condensed nuclei in their central regions, where the stellar population is far denser than in the outlying areas. Here, of course, we are in the favourable position of being able to view these external stellar systems from the outside. It is not possible to resolve such nuclei into individual stars when they are extremely closely packed but there are several other considerations which led the early investigators to believe that this is, indeed, the case. When we come to examine our own Galaxy, the problem is further complicated by the interstellar extinction due to gas and dust which is so great that it is quite impossible for us to see into the galactic centre at optical wavelengths.

The first breakthrough in this direction came some years ago when radio astronomers discovered a very powerful source of radio emission, designated Sagittarius A which, from a study of stellar motions, was found to lie almost exactly at the centre of the Galaxy. Following the discovery by Backlin of a faint infra-red source in almost the same position, Neugebauer and Leighton examined this particular object at various wavelengths in the infra-red, gaining much useful information concerning the distribution of stars close to the nucleus. In addition to a very extended infra-red source which

180

presents an extremely complicated pattern, possibly owing more to variations in the density of obscuring matter in this region than to any intrinsic changes in the original radiation, there appears to be an extremely small point source which, if it is situated at the centre of the Galaxy, cannot be more than a third of a light-year in diameter.

In spite of the small size of this source, the radiation given out from it is equivalent to that of more than a quarter of a million Suns, similar to that of a nova outburst. Very few stars indeed are known which pour out energy at this prodigious rate; S Doradus and the peculiar nova-like variable η Carinae are among them. One may argue that this particular object is not a single star but a very dense cluster. This would certainly overcome the problem of postulating the existence of an almost unique type of stellar object, but we immediately run into another grave difficulty. The crowding of a cluster of about 250,000 stars similar to the Sun into such an incredibly small volume would inevitably result in numerous collisions and the expected lifetime of such a cluster would be far less than the present age of the Galaxy itself. At the present time, therefore, we cannot say with any degree of certainty just what this strange infra-red source is.

The use of infra-red techniques in the future will doubtless provide astronomers with important information on both the structure of our own Galaxy and that of similar galaxies scattered throughout the nearer regions of the observable universe. In particular, such work is now being directed towards a better understanding of the newly-discovered quasars and the peculiar Seyfert galaxies, but more will be said about this in later chapters. Further valuable data will certainly be accumulated to clarify our knowledge of the birth and death of the stars since these two extreme periods in their evolution are more amenable to examination in the infra-red than in the visible.

Ultra-Violet Radiation in Astronomy

At the other end of the visible spectrum, we enter the realm of the ultra-violet, radiations with much shorter wavelengths than those we have just been discussing. The intensity of the ultra-violet radiation that reaches the surface of the Earth is lower than that of the infra-red for a variety of reasons.

The Sun, which provides most of the visible light, is also the source of most of the ultra-violet light within the solar system. Now we must first of all understand how the various radiations are formed within the Sun. Most of the thermonuclear reactions are going on deep in the central core and most of the energy created there is in the form of γ-rays. These are of very short wavelength and extremely energetic. For example, a single quantum of γ-radiation can release several million times as much energy as one of visible light. As we have seen, this tremendous flood of energy works its way outward to the solar surface, first by the process of radiation and then by convection as it nears the photosphere. One of the later satellites put up by the United States, Explorer XI, carried gamma-burst counters on board to measure the intensity of this radiation emitted by the Sun. Fortunately for life on Earth, very little γ-radiation leaves the Sun and quite evidently the prodigious energy released in this form at the centre must become changed during its passage outward through the solar mass.

The only way in which this can be done is by collision with other atoms within the Sun. If a γ-ray hits an atom with sufficient force to knock one of the innermost electrons right out of the atom, radiation in the form of an X-ray is normally emitted. Other types of radiation also occur. If an inner electron is pushed into an outer orbit without being ejected from the atom, then radiation of a somewhat longer wavelength is emitted (ultra-violet light) as the electron returns to its normal orbit. Similarly, less energetic transitions within the atoms lead to emission of visible and then of infra-red light. Radiation of very long wavelengths in the form of radio waves is not normally given out by changes in the electrons of atoms; rather it is emitted when electrons themselves interact with magnetic fields.

We may, therefore, visualize the various types of radiation given out from the surface of the Sun as due to collisions of these kinds, with virtually the whole of the original γ-radiation being transmuted into longer wavelengths as energy works its way out from the centre.

This is not to say that this mechanism is the only one to produce such radiation. Above the photosphere lies the solar atmosphere, which may be conveniently divided into two regions. The lower one, which extends to a height of about 10,000 kilometres above the photosphere, is known as the chromosphere, the higher portion being the corona. When we examine the radiation that is emitted by the solar atmosphere, which is conveniently done by making observations from either rockets or orbiting satellites, we find that as well as visible light the chromosphere emits an appreciable amount of ultra-violet radiation and the corona a high proportion of X-rays.

This is a peculiar result, which we would not expect from a preliminary consideration of the temperature of the solar atmosphere or of the nuclear reactions present. To understand why these particular kinds of radiation arise within the atmosphere we must first

182

examine just what the temperature is in this region of the Sun. Close to the photosphere, within a thickness of some 2,000 kilometres, the temperature is about 5,700°K, not very different from that of the photosphere itself. As we go higher, however, instead of decreasing the temperature commences to rise. At 3,000 kilometres above the photosphere it is slightly higher than 7,000°K. A rise of a further 1,000 kilometres brings the temperature into the region of 25,000°K, while at the outermost limits we have been able to detect, it is probably in excess of 100,000°K. Indeed, it is thought that in some of the outermost regions it may exceed the million-degree mark. Obviously, when we use the term 'temperature' in this context it means something different from what we have used to describe the conditions deep within the interior of the Sun. If the solar atmosphere were radiating at a temperature of close on 1,000,000°K, the entire solar system would be incinerated by this tremendous flood of radiation.

When we say that the corona is at a temperature of 1,000,000°K, we mean the kinetic temperature, which is quite different from the radiation temperature deep within the Sun. Kinetic temperature gives us a measure of the velocity with which the individual atoms are travelling. For example, a single hydrogen atom in interstellar space may be moving with a velocity of several thousand miles per second, giving it a kinetic temperature of 1,000,000° or so, whereas its radiation temperature (the kind with which we are familiar here on Earth whenever we speak of a hot body) would be close to absolute zero. What is important is that these high velocities in the solar atmosphere are capable of bringing about collisions between atoms and other nuclear particles similar to those that are going on inside the Sun. Such collisions result in electrons being knocked from one orbit to another in exactly the same way as before, producing ultra-violet radiation in the lower levels where the kinetic temperature is quite low and more energetic X-rays at the higher altitudes.

Photographs of the Sun taken in ultra-violet light show it to possess a curiously mottled surface and we now know that much of the ultra-violet radiation is associated with sunspots, and particularly with what are known as solar flares. A flare always occurs in association with a sunspot or with the area of the solar surface between two or more spots. It is only a transient thing, forming very rapidly and lasting for only a few minutes; it is probably an electrical phenomenon since the sunspots are known to be regions of high magnetic activity. As well as visual light, a flare emits both ultra-violet radiation and high-energy X-rays which are propagated

183

through space with the speed of light, reaching the Earth in a little over eight minutes.

Whenever these bursts of radiation strike the terrestrial atmosphere they ionize the gases, particularly those of the D-layer, the lowest layer of the ionosphere, rendering it virtually opaque to very long wavelengths but transparent to the short waves used in radio communication. The short waves therefore pass out into space, disrupting communications over large areas of the Earth.

High-altitude rockets and orbiting satellites have also taken photographs of the stars in ultra-violet light, with interesting results. Those stars which have spectral types later than B in the Harvard Sequence scarcely show on these photographs, but the massive and brilliant Wolf–Rayet and Orion stars have such high surface temperatures, in excess of 25,000°K, that most of their radiation comes through in the ultra-violet. The Orion nebula, for example, is an extremely powerful ultra-violet source.

X-Ray Stars

A large number of X-ray photographs of the Sun have now been taken by American Aerobee-Hi rockets and have already taught astronomers a tremendous amount about the structure of the corona where, as we have seen, much of the Sun's X-ray emission originates. Those X-rays generated deep within the solar regions do not escape to the surface at all. Those which are generated by solar flares and in the corona can, of course, escape easily into space and, like the ultra-violet radiation just discussed, also produce ionization in the terrestrial atmosphere, with similar results.

8. Listening to the Universe

For a long time it has been known that there is a background noise present in a radio receiver which, no matter how perfect the components or how far the receiver may be removed from any source of terrestrial radio emission, can never be entirely eliminated. The reasons for this excess of background crackle were extensively investigated during the early days of radio communication and as early as 1926 Jansky in America began experiments in sweeping the sky in the hope of solving the problem of this irreducible noise. Two years of intensive research led him to the conclusion that part of this static was due to Hertzian waves reaching the Earth from outer space. By 1935 he was able to employ aerials which were far more directional in their reception and he discovered that the intensity of the noise increased as the aerials were moved progressively in the direction of the Milky Way and, moreover, that the maximum intensity was produced when they were pointed towards the galactic centre, in the region of Sagittarius.

Jansky's work was all carried out on the 10-metre waveband, but a few years later Reber employed a paraboloid-dish type of receiver which can be classed as the first of the radio telescopes as we know them today. This instrument operated on a somewhat shorter wavelength of 1·87 metres, but the rapid development of the new science of radioastronomy did not really commence until just after the War, following the tremendous amount of research carried out at even shorter wavelengths, particularly in the centimetre range that was the outcome of the work on radar.

Apart from the limited optical window in the electromagnetic spectrum, the radio-frequency band from about 1 centimetre to 70 metres is the only other range of wavelengths which penetrates more of less freely all the way to the terrestrial surface. Of the two, the radio window is the wider and the view of the universe which the radio-astronomer receives is quite different from that of his optical colleague.

Radio Telescopes

There are two main types of radio telescope now in general use. Reber's instrument, as we have seen, was of the paraboloid-dish

185

type and had a diameter of 32 feet. Much larger instruments are now in existence. The Jodrell Bank radio telescope which is of this type, has a diameter of 250 feet, and a second, located at Green Bank, West Virginia, is 300 feet in diameter. These instruments have the distinct advantage that they are steerable and may be used to examine large areas of the heavens.

A much larger paraboloid-dish is that at Arecibo in Puerto Rico, which is operated by Cornell University. Here a large natural formation in the rock has been reshaped, not into a paraboloid shape which would be an extremely difficult engineering problem, but into a spherical shape, which has then been covered with a reflecting surface consisting of steel mesh, the entire bowl being some 1,000 feet across. Unlike those previously mentioned, this is a fixed instrument but by means of a special feed system it is possible to use it to cover an area up to about 20° on either side of the zenith.

A fourth radio telescope of the dish type is now operating in the Goobang Valley in Australia and being of the steerable kind can cover wide regions of the heavens which lie beyond the reach of astronomers in the Northern Hemisphere.

Although the main application of the dish-type radio telescopes is that of picking up radio signals from extra-terrestrial sources, they can also act as radar transmitters and receivers, emitting short-wavelength pulses towards the Moon or the planets. The pulses are reflected by these bodies and picked up again at the source, the time lag giving a direct estimate of the distance. At present, such distances have been determined out as far as Saturn and it seems quite likely that as this technique is further developed radar contact with Pluto may be established in the near future. The axial rotation of a planet may also be measured by the Doppler broadening of the return signal, this method being used particularly in the case of Venus, where it is impossible to see any surface markings by ordinary techniques. The nature of a planetary surface has also been studied from the characteristics of the return signal.

A second type of radio telescope, which may be used for high-resolution work, is that known as an interferometer. This is simply an array of identical dipoles or Yagi antennas, rather like television aerials, usually spaced in the form of a cross, although other con-figurations, for example a semicircle or circle, may be employed. One of the largest of these radio telescopes is situated in Australia at the Molonglo Observatory; this instrument is capable of adjust-ment so that the cross, which has arms about a mile in length, can also be used as a transit instrument.

Observations in Radio-Astronomy

The science of radio-astronomy is of relatively recent origin, but even in this comparatively short space of time so many new and exciting discoveries have been made, the boundaries of the known universe have been pushed back so much farther, that it is difficult to visualize how our knowledge will be augmented by this new technique in future years. It is tempting to speculate what additional information could be obtained if we were able to view the universe in the entire range of wavelengths in the electromagnetic spectrum. Unfortunately, as we have already seen, our atmosphere prevents us from doing this from the terrestrial surface and many astronomers are now developing techniques for making observations from above the atmosphere by means of rocket-borne equipment.

In spite of this inability to observe satisfactorily except at certain specific wavelengths, radio-astronomy has revealed several different sources of radio emission both within and outside the Galaxy. The Sun, the nearest and brightest of all the stars, is the second or third brightest object to the radio-astronomers and powerful radio waves are produced by the Sun. Some of this radiation is closely associated with violent disturbances within the solar atmosphere, particularly the sunspots and prominences, which have been extensively studied by Australian radio physicists. Piddington and Davies have shown from measurements of radio emission from the corona that in the vicinity of sunspots the temperature of the corona is higher than in other regions and, since there are fairly intense magnetic fields also associated with the sunspots, it seems likely that the various particles, which are moving with high velocities, interact with these magnetic fields to produce radio waves. In addition, we often find that the Sun emits extremely powerful bursts of radio waves, which in this case are formed by solar flares, these too being formed in the neighbourhood of sunspots. From the evidence we have, it appears that two jets of particles are ejected from a solar flare – a main burst, travelling at around 1,000 kilometres per second which is succeeded by a less intense burst travelling at close to the velocity of light.

High-Temperature Galactic Hydrogen

When we discussed the evolution of the stars in Chapter 4, we saw how certain regions of the Galaxy contain a profusion of young blue giants, closely associated with the large gas clouds out of which they have been born comparatively recently. One of the best known of these regions is the Orion Nebula which is literally teeming with such stars, many of which are variable. Other areas, which are less precisely defined, exist along the plane of the Milky Way.

187

In the early 1950s, Ryle and Scheuer detected a general emission of radio waves from the regions bordering the galatic plane, these emanating from clouds of hot hydrogen. Somewhat more recently, radio physicists at the Naval Research Laboratory in Washington, led by Haddock, Mayer, and Sloanaker, re-examined these regions at a wavelength of 9·4 centimetres and were able to delineate two individual clouds, these being the Trifid nebula, Messier 20 in Sagittarius, and the Orion nebula (see Plate IX).

Why should a large cloud of hot hydrogen gas emit radio waves in addition to ordinary light? The answer appears to lie in the nature of the stars embedded within such clouds. These are almost all stars that fall high on the main sequence, giant stars with high surface temperatures of the order of 25,000°K, and such stars emit a large proportion of their radiation in the far ultra-violet. Now one characteristic of ultra-violet radiation is its ability to ionize neutral atoms; consequently much of the radiation at these short wavelengths is absorbed by the surrounding hydrogen gas, serving to break the bond between the proton and the electron.

Owing to the large numbers of such stars within these clouds, much of the hot hydrogen is composed of free protons and electrons which, because of the tenuous nature of the gas, are able to move for relatively large distances before colliding. Now when a proton and an electron collide, two things may happen. If they approach each other sufficiently closely at the correct velocities, they will recombine to form an atom of neutral hydrogen and in the process visible light will be emitted. If, on the other hand, the conditions are not right for recombination, part of the energy of collision is transformed into radiation covering a very wide range of wavelengths; visible light, ultra-violet, infra-red and, as Henyey and Keenan have shown, radio emission.

One important characteristic of these vast clouds of hot hydrogen within the Galaxy is that, since radio waves are not affected by obscuring matter to the same extent as visible light, they can be observed at far greater distances in the direction of the galactic centre. In Chapter 12, when we discuss the quasars, we shall see that such infra-red and radio emission can be observed directly at the centre of the Galaxy itself, leading to very important conclusions about the state of affairs existing there.

Discrete Galactic Radio Sources

In addition to the huge clouds of high-temperature hydrogen, other radio sources exist within the Galaxy, including those containing various molecules, which will be considered separately in the following chapter, and a large number of discrete sources; these are extremely powerful radio emitters, some of them being a million times more powerful than any of the hydrogen clouds.

This tremendous output of radio energy cannot be explained on the basis of interactiing protons and electrons and we are forced to examine other mechanisms that may be sufficiently energetic to produce this radiation. First, however, let us look at the physical characteristics of these sources. One such source is the Crab nebula in Taurus, which is the remains of a supernova that flared up in 1054. Another is an exceptionally powerful source in Cassiopeia which, although not as outstanding visually as the Crab nebula, appears on long-exposure photographs as a series of fine filaments. These filaments have been shown by Baade and Minkowski to be in a constant state of motion. Some of them, those which are more clearly defined on the photographs, are moving toward us with velocities up to 2000 kilometres per second while others, more diffuse and not readily visible, have even higher velocities, around 4,500 kilometres per second, with certain filaments often showing marked changes of velocity throughout their length. A third intense radio source in the constellation Cygnus is extremely faint photographically and was identified as an extra-galactic object in 1954.

Now why should such filaments of gas, even if moving with these high velocities, emit radio waves? A plausible answer has been given by Alfven and Herlofson, who suggest that whenever the material in these moving filaments collides violently the natural magnetic fields associated with these clouds of gas are compressed, with a resultant increase in their intensity. An added and important result will also be that such magnetic collisions will accelerate free electrons to very high velocities, with kinetic energies which are at the lower end of the cosmic-ray range. Particles of such high velocity are quite capable of producing radio emission when they interact with the general magnetic fields present within these moving filaments.

Apart from the fact that the Crab nebula is known to be a supernova remnant, we have said little about the origin of these discrete galactic radio sources. It now appears quite probable that most, if not all, of the radio sources of this kind are the remains of supernova explosions. Both Tycho's supernova of 1572 in Cassiopeia and that of 1604 in Ophiuchus, have been identified with radio sources. Certainly, a supernova will impart the very high velocities required by the foregoing theory to the expanding shell of gas. Moreover, as we shall see in Chapter 13, the newly discovered pulsars, which are extremely peculiar radio sources, are closely associated with supernova remnants.

189

In addition to the usually well-defined sources just mentioned, other regions of radio emission exist that are associated with the Galaxy. For example, a more general emitting region has been found which appears to stretch beyond the plane of the Galaxy. So far no discrete sources are detectable within this vast, somewhat diffuse, region and here it seems probable that high-energy electrons are produced within certain areas such as the very powerful source in Cassiopeia and are able to escape into a general medium surrounding the Galaxy. Such electrons, by virtue of their high velocities, may themselves be able to produce radio emission, possibly by inter-action with the general magnetic field of the Galaxy itself.

The central regions of the Galaxy also emit intense radio waves, but here the situation is somewhat different. As we shall see in a later chapter, it has been suggested that here there is a dead quasar which is the source of this emission, and more will be said on this subject when we come to discuss the quasars.

Extragalactic Radio Sources

It seems only reasonable to assume that, just as our own Galaxy emits radio waves, other galaxies will do likewise. The first survey of such radio sources was made by Ryle, who was able to identify about sixty of them, and observation showed that they were distribu-ted almost equally in all directions. This uniformity indicated that they lay beyond the bounds of our own Galaxy and were not merely sources right on the periphery since if this were the case we would expect them to be concentrated along the Milky Way. The proposal that most of these sources were extragalactic, possibly other galaxies similar to our own, was first put forward by Gold and the evidence we now have, including the discovery by Hanbury-Brown and Hazard that the Andromeda spiral is a weak emitter of radio waves, is strongly in favour of this interpretation.

Originally, such sources were termed 'radio-stars', but since the extragalactic nature of these objects was established this designation has been generally abandoned. Once it became possible to measure the intensity of the radio emission from the nearer galaxies, it was found that in the case of the elliptical galaxies no correlation exists between the intensity of the visible radiation and that of the radio emission, but for the large spirals there is a simple relation between these two parameters, the intensity of the radio emission keeping in step with that of the visible light. Now why should this be so?

The answer appears to lie in the intrinsic nature of these two types of galaxy. The elliptical galaxies all contain stars that are of Baade's Population II, these being the oldest stars that were

190

formed out of the original gas. Similar stars are also found in the large spirals, but only in the nuclei and the globular clusters; the stars present in the spiral arms themselves (and those that are virtually absent from the elliptical galaxies) are of Population I. Now Population I stars have a wide range of ages, varying from stars only a few million years old to others, like the Sun, that are almost as old as the Population II stars. The young Population I stars are found high on the main sequence and, as we have already seen, are responsible for the radio emission from clouds of hot hydrogen within our own Galaxy. In addition, this proves that the spiral arms also contain a large amount of gas still available for star formation and such gas is a necessary constituent for this type of radio emission, since once the component particles, the electrons, are accelerated to high velocities they will interact with magnetic fields present within the galaxies. Having little or no gas available, the elliptical galaxies will not be expected to produce such radio emission on the same scale. The radio emission from the spiral galaxies will therefore be proportional to the number of bright young stars of Population I and therefore to the integrated light from the spiral arms, whereas in the elliptical galaxies this simple relation will not hold since the number of these stars is necessarily small and most of the gas and dust available for star formation has already been used up or dissipated.

Colliding Galaxies

We must now consider a very peculiar situation. From what has been said, we might expect the large spirals to be the strongest radio-emitters among the galaxies, whereas observation shows that, in reality, the most powerful radio galaxies are ellipticals and certain of them appear to be very peculiar objects. One such source is shown in Plate XVI; another is a pair of galaxies NGC 4567 and 4568 in Virgo. Here the intensity of the radio emission is somewhat greater than that of the visible light.

Originally it was considered that such strong radio sources were colliding galaxies, particularly since such collisions would produce radio emission by interaction of the gas and dust clouds. This theory has been largely abandoned now since the scale of the universe is believed to be such that collisions are too rare to account for the high number of such radio sources. The structure of radio galaxies have also been found to be double with a large separation of the two components.

191

9. Molecules in Interstellar Space

In the preceding chapter we saw how radio emission has now been added to those radiations which may be used by the astronomer in his study of the universe and especially how radio-astronomy has enabled us to probe farther into space than is possible with optical telescopes. We shall here examine in more detail some of these radio emissions which emanate from regions within the Galaxy, with particular reference to the way in which they have demonstrated the presence within interstellar space of various molecules, many of which are present as large, diffuse clouds in certain well-defined galactic regions.

The information we obtain from the study of these clouds has proved of great value in furthering our understanding both of the structure of the Galaxy and of the various physical processes going on within it. It will readily be appreciated that although we can make fairly accurate star counts over the whole of the sky and thereby obtain a good idea of their distribution, this will provide us only with a picture of the local region of the Galaxy lying in the neighbourhood of the Sun. The reason for this is not far to seek. Whenever we use the optical telescopes to observe the stars in the direction of the galatical centre we are hampered not only by the tremendous number of stars in this area but also by the large regions of obscuring dust which prevent us from probing very far in this direction.

From observation of other galaxies we can reasonably presume that our own possesses a spiral structure but, so long as we could only use optical methods, proving this assumption was an extremely difficult matter. The distribution of stars in the region around the Sun certainly indicated the presence of three spiral arms but unfortunately only short segments of these arms could be satisfactorily mapped by this optical method. What was required was some means of mapping the huge gas clouds – invisible either visually or photographically – that lie along and between the spiral arms. It is an obvious suggestion that, since hydrogen is by far the most abundant element in the Galaxy as a whole, there ought to be some means of detecting its presence, not in the stars themselves, but in the form of vast

192

clouds among the stars. Where there are very hot, supergiant stars embedded within such clouds of hydrogen, the intense ultra-violet radiation from these stars results in the heating of the surrounding gas, thereby rendering it visible as, for example, in the Orion nebula shown in Plate IX.

When it comes to detecting cool clouds of neutral hydrogen, however, the problem is far more difficult. Such clouds emit no visible light and although, theoretically, such vast gaseous regions should exist within the Galaxy, no means were available for observing them. Spectroscopy would appear to be the most useful tool for their detection but unfortunately as far as hydrogen is concerned only the Balmer series of lines is observable in the visible region of the spectrum and this series corresponds to only a very restricted range of excitation of between 10 and 12 electron volts. The two really important series of spectral lines of hydrogen are extremely difficult to observe; these are the Lyman series, lying completely in the unobservable region of the ultra-violet, and the Paschen series, in the far infra-red, both of these regions being masked by absorption due to the terrestrial atmosphere.

In 1944, however, it was suggested by van de Hulst, on purely theoretical grounds, that radio waves should be emitted by neutral hydrogen at a low temperature at a wavelength of 21 centimetres. Fortunately for the radio-astronomers, this particular wavelength is conveniently placed for observation. Now why should a neutral hydrogen atom emit radiation at 21 centimetres? To understand this we must examine the hydrogen atom in some detail. As we have already seen, it consists of a proton and a single orbital electron and both of these behave as tiny magnets. The hydrogen atom may therefore exist in two distinct states, one in which the parallel axes of the proton and electron lie in the same direction and another in which they are aligned in opposite magnetic directions. Whenever the axis of the electron flips over from one direction to the other, the 21 centimetre radio wave, corresponding to a tiny amount of energy, is emitted. Although this emission of energy occurs spontaneously, it does so on the average only once every 11 million years for any particular atom. In other words, this is a forbidden radiation. Under ordinary conditions. such as those prevailing inside a star, this radiation, although emitted at the same rate per hydrogen atom as in a gas cloud, is swamped by other emission and absorption processes which here occur with much greater probability. The one fundamental condition that must be fulfilled if we are to be able to detect this radio emission is that the density of the hydrogen cloud must be extremely low. Such a condition is realized only when the hydrogen exists spread out as in interstellar space. Here there are only about two hydrogen atoms to each cubic centimetre; in spite of this low density, however, interstellar space is so vast and these clouds so huge that the total number of hydrogen atoms is very great.

Distribution of Cool Interstellar Hydrogen

Seven years after van de Hulst's theoretical prediction, the detection of the 21-centimetre emission line was announced almost simultaneously by three independent groups working in America, Holland and Australia. Since this initial discovery, a tremendous amount of work has been carried out using the largest radio telescopes to delineate the boundaries of these large clouds of neutral hydrogen and also to determine their distances, for here we possess an extremely powerful tool for mapping the spiral structure of the Galaxy.

The major difficulty in this work, of course, is that we are rather in the position of someone trying to obtain an accurate picture of a maze while being unable to move from one particular point inside it. Were we situated inside one of the Magellanic Clouds or the spiral nebula in Andromeda, for example, it would be a relatively simple matter for us to view the spiral structure of our own Galaxy. As it is, we must have recourse to other methods and the radio-astronomical observations of the 21-centimetre line have recently provided us with an excellent and quite accurate map of the local spiral structure of the Galaxy (Fig. 40).

There remains the problem of determining the distances of these hydrogen clouds within the Galaxy. This is done most conveniently by taking account of the rotation of the Galaxy itself. The hydrogen clouds are not stationary but partake of the rotational motion of the Galaxy, having motions that are dependent upon their distance from the centre. One consequence of the type of rotating with which we are dealing is that angular velocities are inversely proportional to the distance from the galactic centre, and from the pioneering work of Lindblad and Oort we know that the Sun, which is just over 8,000 parsecs from the centre, has a rotational velocity of 135 miles per second. Such velocities are sufficiently high to produce a detectable Doppler shift in the 21-centimetre line and from the velocities determined for individual hydrogen clouds we are able, after taking into account our own velocity, to deduce their distances from the centre of the Galaxy and thereby their relative positions within the galactic arms.

A logical extension of this method is the detection of cool, neutral hydrogen clouds within other galaxies. The first application of this

194

Fig. 40 Diagrammatic map of the spiral structure in the local region of the Galaxy as determined from infra-red observations.

new technique to external galaxies was made by Kerr and Hyndman, who examined the two Magellanic Clouds with some very interesting results. Both of these nearby irregular galaxies contain fairly large quantities of neutral hydrogen, and from the Doppler shifts that have been measured it is apparent that not only do the Magellanic Clouds possess motions within themselves but they are also revolving about each other, rather like a binary star system although on a vastly larger scale.

More recently, this method has demonstrated that certain compact groups of galaxies exert a very definite and powerful gravitational effect upon one another. This had been suspected for several years since in certain cases it is possible to see faint, nebulous streamers of gas which bridge two or more galaxies. Whether there is also a magnetic field connecting such galaxies an open question at the present time although this appears likely.

195

Radio Emission from Water and Ammonia in the Galaxy

The discovery of radio emission from neutral hydrogen within the Galaxy inevitably prompted investigations into the possibility of detecting other neutral molecules in galactic space. Within the past few years, both neutral water and ammonia clouds have been found, particularly by Townes at Berkeley and Cheung and his colleagues using the 20-foot millimetre-wave radio telescope at the Hat Creek Observatory. Both ammonia and water vapour appear to be fairly widespread within cool dust clouds in the Galaxy, and several of these have been located in the direction of the galactic centre in Sagittarius. The temperatures of these clouds appear to lie between 35 and 150°K.

Four spectral lines due to the hydroxyl radical of water appear near 18 centimetres, compared with 21 centimetres for neutral hydrogen. That of ammonia is at 1·25 centimetres and both formaldehyde and carbon monoxide have been discovered in the microwave band and molecular hydrogen in the rocket ultra-violet. Since it is thought that star formation takes place within the cool dust clouds in the Galaxy, especially those where water vapour is also present, the detection of ammonia in these clouds is of particular importance, since it may tell astronomers what the role of nitrogen might be in the formation of stars. Just how the ammonia is formed initially is still a problem, although it appears quite probable that both nitrogen and hydrogen are adsorbed on the surface of the small dust grains within these cool clouds where chemical combination of the two gases takes place. Once formed, the ammonia may be liberated from the surface of the grains by several well-known processes. It may, for example, simply sublime into space, or it may be liberated by bombardment with other particles or by the action of ultra-violet radiation.

Undoubtedly, radio-astronomers will be scanning other dust clouds for evidence of these various molecules within the Galaxy and these investigations will prove of immense importance in furthering our understanding of the structure of the Galaxy.

10. The Galaxies

One of the great revelations in astronomy during the present century is undoubtedly the discovery that beyond the Milky Way, beyond the globular clusters which form a gigantic halo around the nucleus of our own Galaxy of a hundred thousand million stars, there exist other galaxies, many of which are comparable in size and structure to ours. A mere handful of these are visible to the naked eye and were known to the ancients. The Andromeda nebula, for example, was classed as a star by Al Sûfi in 964 and, although there are no written records available, the two Magellanic Clouds must also have been known to the inhabitants of the Southern Hemisphere long before they were discovered by the first European explorers.

With the invention of the telescope, the number of galaxies known increased enormously, and as early as the mid-eighteenth century Messier catalogued 103 such nebulae. He did little, if anything, however, to investigate their true nature since he was primarily concerned with discovering comets, which have a similar appearance when some distance from the Sun, and merely compiled his catalogue in order to be able to disregard these nebulous clouds among the stars when comet hunting.

At first, all of the nebulae were classed together since no means existed of distinguishing between those that are merely vast clouds of glowing gas within our own stellar system and others that might consist of individual stars but so distant that they appear as little more than faint, hazy patches of light. The suggestion that many of these objects might be huge systems of stars lying beyond the boundaries of the Milky Way – 'island-universes' – was first made by Kant and Swedenborg who were both philosophers rather than astronomers. Such a belief was also held by Sir William Herschel who compiled the first really comprehensive catalogue of the nebulae. Unfortunately, the vast majority of the objects included in his catalogue were gaseous nebulae and during the nineteenth century, when spectroscopy was first applied to the study of these objects, the discovery that most were merely clouds of glowing gas led to the virtual abandonment of the idea of extragalactic nebulae.

There were some, of course, whose spectra were entirely different and resembled those of ordinary stars in that they showed dark

197

absorption lines instead of a few bright emission lines, as we would expect from an incandescent gas. One further point in favour of the island-universe hypothesis came when the spiral nature of many of these nebulae was recognized as larger telescopes came into operation. By 1908, with almost 13,000 galaxies discovered, about three-quarters of these exhibited a recognizable spiral structure.

At this time, too, Ritchey, using the 60-inch reflector at Mount Wilson, was able to resolve a few of the nearby galaxies into what seemed to be individual stars. The images produced on the photographic plates were not, however, truly stellar but had a curiously fuzzy appearance so that it was impossible to state with absolute certainty that they were stars and not small star-like objects, possibly tiny gas clouds. We must remember, too, that at this time the distances to the nebulae were unknown; there was also the argument, convincing at that period, that since no nebulae were observed along the plane of the Milky Way they must necessarily be associated with it. We now know that this argument is erroneous in that the dust along the galactic plane effectively obscures those nebulae that lie in this direction.

In 1912 Miss Leavitt had found that the Cepheid variables present in the Large Magellanic Cloud provided a relationship between their period and intrinsic luminosity and it was therefore possible to use these stars to obtain measurements of distance. This was the first direct observational evidence that other systems of stars lay beyond the confines of our own Galaxy. In itself, however, it was not conclusive evidence that the other spiral nebulae were the island universes envisaged by Kant. Not only are the two Magellanic Clouds very irregular in outline, being loose collections of stars with little, if any, spiral structure but, at a distance of only about 180,000 light-years, they are really mere satellites of our own Galaxy.

Then, in 1917, Ritchey discovered a nova in the galaxy NGC 6946 in Cepheus on a plate taken with the 60-inch reflector. From its apparent magnitude at maximum, and assuming that it was a typical nova, he estimated its distance as several hundred thousand light-years, well beyond the boundaries of the Galaxy. A subsequent search through the photographic records turned up several more which appeared to be equally distant. In spite of this photographic evidence, the astronomical world remained sharply divided on the question of the extragalactic nature of the spiral nebulae for almost seven years.

Then, late in 1924, the 100-inch telescope at Mount Wilson was brought into commission and Hubble was able to resolve star images in three spiral galaxies – M 31 in Andromeda, NGC 6822, and M 33.

198

This in itself was an important step forward, but what proved to be more exciting was that several of these stars were Cepheid variables and could therefore be used as distance indicators. Refinement of the early measurements, including the recognition that the RR Lyrae stars and the classical Cepheids are two distinct types of variable star, has shown that the Andromeda spiral lies at a distance of 2,200,000 light-years and that in size and stellar content it is comparable with our own Galaxy.

Having established the nature of the galaxies, the 100-inch telescope and later the 200-inch at Palomar were used to push back the frontiers of distance to their present limits. We now know that the number of galaxies runs into thousands of millions, that there are as many of them as there are stars in our own Galaxy.

The Cepheid method of distance determination has been used successfully out to about 3 million light-years. Beyond this limit such stars are so faint as to be virtually indiscernible. Up to a distance of around 30 million light-years, it is possible to use the very bright, blue supergiants that are present in the spiral arms of the galaxies, and intergalactic distances may be measured a little further than this when we use the supernovae, although unfortunately these are of much rarer occurrence than the other stars. If we wish to measure even greater distances than these, we have to fall back upon the method, used particularly by Hubble and Humason, of utilizing the average brightness of whole galaxies, that is, those that correspond to the types whose distances have already been measured with a fair degree of accuracy by one of the methods just outlined. As will be readily appreciated, the accuracy of the figures obtained decreases with increasing distance until, at the very limit to which we can observe, the results are continually being modified as more refined techniques come into use, often by as much as 30 to 40 per cent.

Classification of Galaxies

As far as mere size goes, most of the galaxies are somewhat smaller than our own. In shape they vary widely from loose, highly irregular clusters of stars to tightly-wound spirals, which are viewed in the telescope from all possible angles, some seen full-face, others edge-on and many at intermediate inclinations to our line of sight. As a result, we must make full allowance for perspective if we are to make any proper and accurate classification. Hubble, who made the first survey with the 100-inch reflector, classified them into three main types – the irregulars, the spirals and the ellipticals.

The irregulars are all ill-defined; although there may be a hint of

199

spiral structure at times, in most cases this is completely absent, as is also a nucleus. The best-known examples are the Magellanic Clouds. When we examine the stars present in the irregular galaxies, we find that they are predominantly hot, blue stars of Population I, very high up on the main sequence. In addition, these galaxies contain an abundance of gas and dust, an observation we would expect from the presence of young stars. Several of the irregulars are also characterized by the presence of a bar-like structure and this, together with the fact that there is plenty of interstellar dust and gas available for star formation, has led several astronomers to the conclusion that these galaxies may be the precursors of the second large class, the spirals.

Whereas the irregulars form only about 2 per cent of the total number of known galaxies whose structures have been determined, almost 80 per cent are spirals. Since these latter galaxies have a more or less well-defined appearance, it is possible to divide them into a sequence, or rather two sequences. the normal spirals are classified as: (1) Sa spirals, in which the nucleus is very large and pronounced and the arms are under-developed, in general tightly bound to the nucleus; (2) Sb spirals, which form an intermediate group with both the nucleus and the arms approximately equally developed; and (3) Sc spirals, which have a very extensive and loose spiral arm structure and only a small nucleus.

The second sequence of spirals has a curiously barred structure with the arms beginning at either end of a bar that appears to extend all the way across the nucleus. These are similarly designated as types SBa, SBb and SBc galaxies, their classification bearing a similar connotation depending upon the relative importance of the nucleus and spiral arms. Certain of these barred spirals also show indications of a secondary spiral structure within the nucleus, almost like a set of inner arms. So far, no explanation of this peculiar effect has been given. Examination of the spiral class as a whole in both red and blue light reveals the presence of old Population II stars in their dense nuclei and in the globular clusters that form a halo around the nuclei. Within the spiral arms are the younger Population I stars, many of which are probably still being formed from the gas and dust clouds which preponderate there.

The third class of galaxies, which makes up about 17 per cent of the total, are the ellipticals. These range from globular objects through various ellipsoidal configurations to a limiting lenticular form in which the ratio of the major to the minor axes is about 3 : 1. As with the spirals, it is possible to arrange the elliptical galaxies into a fairly regular sequence from those of type E0, which are virtually spherical, to type E7, which are elongated ellipsoids, examples of all the intermediate classes being known. The sequence ends at E7, since those that would be E8 or higher are almost certainly spiral galaxies seen edge-on; although it is possible that true elliptical galaxies of type E8 may occur, they have not yet been observed and must be exceedingly rare. Ellipticity is defined as $(a-b)\ a$, where a and b are the major and minor diameters respectively; to place any particular elliptical galaxy within the sequence the ellipticity is estimated to one decimal place and the decimal point is omitted in the numerical designation. For example, a galaxy in which the major axis is twice the minor axis would have an ellipticity of $(2-1) - 2$ is 0·5 and would be called an E5 galaxy.

Most of these galaxies show a fairly bright nucleus, with the luminosity decreasing steadily towards ill-defined edges, and accordingly there are two general features that may be used to classify them. First, there are contour shapes, or more precisely the isophotal contours (lines of equal luminosity), and second the brightness gradients. The latter are extremely difficult to measure with any degree of accuracy and are little used for classification.

Even in the case of the isophotal contours, which are readily obtained from photographs, there is an unavoidable difficulty in that they may be said to represent a true picture of an elliptical galaxy only when allowance is made for the orientation, since we see these objects merely as projections against the intergalactic background. Only in the case of the E6 and E7 ellipticals when seen edge-on do the isophotal contours provide a true representation of the galaxy. In spite of this serious difficulty it has been shown by statistical analysis of a large number of these objects that they do range from spherical to lenticular ellipticals, the former being much less frequent than the higher members of the series.

The stars that make up the elliptical galaxies are predominantly the old Population II stars – chiefly red giants and several classes of variable star. The relative dearth of young Population I stars is readily explainable by the comparative absence of star-forming gas and dust.

Distribution of Galaxies

The old idea that was held up to about thirty years ago that the majority of the galaxies are entirely separate from one another, with scarcely any link between them, has now been abandoned. Almost all of them exist in huge clusters, presumably bound together by strong ties of gravity and in those few cases where isolated galaxies have been discovered it now seems that these objects have been ejected from the large groups rather than having been formed in this manner. We have already seen, for example, how the two Magellanic Clouds are satellites of our own stellar system and radio measurements have shown conclusively that these two irregular star systems are revolving about a common centre of gravity and in all probability about our Galaxy too. Other instances of double galaxies are known that are in revolution about each other, and there are cases in which a faint spiral arm connects two close galaxies, such connecting bridges providing us with some evidence not only of gravitational but also of magnetic fields operating between the two systems. From what we know generally of these magnetic fields between galaxies, it appears that they may also be fairly powerful emitters of radio waves.

Apart from the two Magellanic Clouds, there are a further sixteen galaxies known which, together with our own, form what is known as the Local Group. The centre of the Group is somewhere between our Galaxy and the Andromeda spiral, all of the other members lying in a vast sphere with a radius of about 2 million light-years. At this point, it is important to realize that the Local Group is a gravitationally-bound cluster and in this sense we may regard these objects as a single, physically connected unit, although a very small one as such clusters go. One result of this is that the individual members are found scattered in all directions, showing no tendency to group together in any particular region of the sky. For example, one of the nearest is a small galaxy in Draco, while another lies in the southern constellation of Fornax. Although only nineteen members are known at present, it is just possible that there may be a few more belonging to the Local Group which cannot be seen simply because they happen to lie in the direction of the Milky Way, where they are hidden by the obscuring dust clouds in the vicinity of the galactic plane.

Other clusters of galaxies are known that contain many thousands of members. One of the nearest is the Virgo cluster, at a distance of about 50 million light-years. Beyond this lie the Ursa Major cluster at some 650 million light-years, the Corona Borealis cluster at 945 million light-years, the Boötes cluster at 1,700 million light-years and further still, in Hydra, a group of extremely faint galaxies at a mean distance of 2,700 million light-years.

Beyond these limits lie other clusters which are so faint that, although they can be detected photographically, the Doppler shift in their spectra cannot be measured with any real degree of accuracy and estimates of their distance are necessarily crude and subject to large inaccuracies. One very odd galaxy which has been discovered

202

by the radio-astronomers, known as 3C 295, has been shown by Minkowski to be receding from us with the incredible velocity of 67,000 miles per second, placing it at about 4,800 million light-years.

The galaxies shown in Plate XVII, which lie at the limit of the 200-inch telescope, are thought to be receding from us at about two-thirds the speed of light.

Many of the individual galaxies that go to make up the large and distant clusters can be identified only by their ill-defined outlines, and it is certain that large numbers exist which cannot be seen even on long-exposure photographs. Galaxies such as the Magellanic Clouds, for example, are intrinsically small and faint so the total number in each cluster must be far greater than that estimated directly from photographic plates. Our Local Group is therefore a veritable minnow in a sea of whales.

Evolution of Galaxies

In Chapter 4 we discussed the evolution of the stars and saw how most of them may be fitted into a well-defined scheme, with various changes taking place according to the nuclear reactions going on in their interiors. Once the various types of nebula were recognized, it was only natural that astronomers should attempt to find some evolutionary scheme connecting the three main classes. Any hypothesis we may put forward will, at best, be only a tentative one but there are certain clues which point the way towards a sequence from one type of galaxy to another, and these form the basis for an evolutionary path that seems quite plausible in the light of our present knowledge.

Since in proportion to their size the irregulars contain the most gas and dust – the material out of which the stars form – it is logical to assume that these represent the first stage in the development of the galaxies. Here we also find very few red stars of Population II, by far the greater number being young blue giants. In some of the irregulars, the amount of star-forming material outweighs the number of stars themselves, indicating that here we have plenty of matter capable of continuing star formation for a very long time. One apparent stumbling block to this idea that the irregulars are the starting point in the evolution sequence is the question of mass; most of these galaxies are appreciably less massive than the huge spirals. At first sight it is difficult to see why this should be the case, since there should be little, if any, mass loss during the transition from an irregular galaxy to one of the spiral type. One way out of this dilemma is to assume that, as with the stars, the rate of evolution will be much more rapid for the very massive irregulars than for those having relatively little mass. If we accept this possibility, then the absence of the very massive irregulars may be because these galaxies have already passed through the various transitional stages into spirals. The more slowly developing irregulars of low mass have not yet had sufficient time to make this transition. If this is so, then the Magellanic Clouds may be expected to develop into small spiral galaxies at some time in the future. It may be significant in this context to note that, whereas the Large Magellanic Cloud contains predominantly Population I stars, the Small Cloud contains not only stars of both populations but also some globular clusters, which are a feature of the spiral galaxies.

When we come to the spiral and elliptical galaxies, we are on a little more certain ground. The spiral galaxy in Andromeda shown in Plate X, has a nucleus that is very similar in appearance to an elliptical galaxy, and this is true for most of the spirals. The difference between the two types therefore appears to be that whereas the spirals possess an outer disc that forms into the characteristic spiral structure, the ellipticals do not. Another feature, which is equally important in our understanding of the relationship between these two types, is that the spirals still contain appreciable quantities of gas and dust, which are absent in the ellipticals. In the latter, therefore, star formation has virtually ceased.

As a starting point, let us assume that every galaxy has the potential to become a spiral; that is to say, it possesses a nucleus (which may be very small and ill-defined as in the case of the irregulars) and a residue of gas and dust that has not formed into stars and is available to develop into an outer disc. In our own Galaxy, for example, observations made by the radio-astronomers have shown the presence of large hydrogen clouds which appear to be moving outward from the centre towards the spiral arms. The next logical step is to examine the means whereby the gas and dust may be removed from a galaxy before it has had time to evolve an outer disc, thereby resulting in the formation of an elliptical rather than a spiral galaxy.

If the gas within the central regions of a galaxy is all utilized in the formation of stars, then the outward flow of such gas towards the outer rim is either greatly reduced or ceases altogether. Such a situation could lead to a reversal of the gas flow back towards the hub where it would then be consumed in the same way as before. The end result will be a galaxy with characteristics intermediate between those of a spiral and an elliptical one. Such galaxies, although rare, are known and have been designated the S0 galaxies.

204

An alternative means of removing the gas and dust, suggested by Baade and Spitzer, is collisions among the galaxies. Since there is a very strong tendency for galaxies to form into clusters, it is quite possible for collisions to occur from time to time. Now what would happed if two galaxies were to collide? In reality, the effect would not be as drastic as would appear at first sight. There would be few collisions of one star with another since interstellar distances, even in the most densely-packed central regions, are so vast. The dust and gas clouds, however, are an entirely different matter. Here there will be intermingling on a tremendous scale with a subsequent rise in temperature. A major effect of this high temperature will be that much of the gas may be boiled off from the colliding galaxies and completely lost.

A lot will now depend upon when such a collision occurs. If it happens before there is any development of an outer disc, then a typical elliptical galaxy will result. If, on the other hand, it occurs after the formation of outer arms, then an intermediate spiral galaxy of type S0 will form.

Is there any way by which we can say which of these two mechanisms is the more likely? As regards the former method, the only direct observational evidence we have is the existence of the S0 galaxies but, as we have just seen, these may also be produced by the second method. There is, however, one particular piece of observational evidence in favour of the latter. If the collision of two galaxies strips them of gas and dust leading to evolution into elliptical rather than spiral galaxies then, if we choose a cluster that is sufficiently densely packed with galaxies for the frequency of collisions to be fairly high, we would expect to find more elliptical galaxies in proportion to the spirals than in looser clusters, and this does indeed seem to be the case. Furthermore, photographs show that the relative number of ellipticals is greater near the centres of densely packed clusters than out towards the periphery, again in good agreement with this hypothesis.

Peculiar Galaxies

In addition to the run-of-the-mill spirals, irregulars and ellipticals, certain extragalactic nebulae have been found that are extremely peculiar in several respects and which cannot be satisfactorily fitted into the evolutionary scheme just outlined. Many of these are powerful radio transmitters and also look distinctly odd on long-exposure photographs. Plate XVI shows one of these peculiar galaxies in the constellation Centaurus. The idea was originally put forward, as we have seen, that these might be two galaxies in the process of collision, the radio waves being produced by the violent agitation of the inter-mingling gas clouds. There seems little doubt that interaction bet-ween gas and dust clouds on this stupendous scale will produce radio emission; the problem to be examined is whether it will produce such emission on a sufficiently large scale to account for the observed intensity. Another difficulty is that the number of such radio sources appears too great to be explained away simply on the basis of collid-ing galaxies.

Burbidge has suggested that these peculiar objects are either very massive or very old galaxies and that a chain reaction between large supernovae in their nuclei results in a tremendous explosion of inconceivable violence. In other words, within their nuclei there are large numbers of stars having helium cores in excess of Chandra-sekhar's limit (1·44 times the mass of the Sun) which are sufficiently close for a few of these exploding stars to trigger off further explo-sions. Some evidence of the existence of explosions on this vast scale has been obtained quite recently.

Lynds and Sandage have found evidence of a gigantic explosion in the nucleus of M 82 which appears to be on the scale envisaged here. In addition, the Large Magellanic Cloud contains a strange pheno-menon, clearly visible in photographs taken in ultra-violet light, that Shapley has named Constellation III. The region is a huge arc made up of blue stars of very high luminosity within which there is very little neutral hydrogen. Constellation III bears a striking re-semblance to a gigantic shell formed by an explosion of incredible proportions which occurred during the early life of the galaxy. Westerlund and Mathewson have estimated the mass of the object which produced this explosion as close on 100,000 solar masses.

Before this discovery was made, Shklovsky had suggested that such explosions occur during the early evolutionary stages of a galaxy, producing the heavy elements that are found within the stars which form later. On the basis of Shklovsky's hypothesis, the super-novae that occur shortly after the formation of a galaxy are far more frequent and powerful than those during its later life history, since they were much more massive, having just formed out of the primeval gas cloud, and contained a much higher reserve of hydrogen. Similar expanding spherical shells have been discovered in a number of other galaxies.

Formation of Galaxies

So far in this discussion we have been skirting around one import-ant problem, namely how the galaxies condense in the first place out of the supragalactic cloud. It will be noticed here that we are speaking not of the condensation of a single galaxy, but of a whole shower of them forming from the same cloud. The fact that the galaxies are found in huge clusters makes it necessary that we consider formation on such a large scale. Fortunately there are several hints as to the nature of the process of galaxy formation. As we already know, the Population II stars are those first formed from the galactic gas and, from the numbers of these stars present both in the nuclei and the globular clusters that form the haloes around the galaxies (haloes which may be as much as three times the size of the galaxies themselves), it is clear that, unlike the later Population I stars, those of Population II are produced in millions during what may be regarded as a burst of star formation. Quite clearly there is a very close correlation between the formation of the galaxies and that of the Population II stars.

One important feature of the supragalactic cloud which influences how it will evolve is its initial temperature. Now calculation of the various factors involved indicates that the temperatures of these clouds fall within two fairly narrow ranges, either 10,000 to 25,000°K or between 200,000 and 1,000,000°K. What happens to the cloud as it begins to condense depends largely upon its initial temperature.

Since we are considering a mass of gas before any stars have formed, the only constituent we need take into account is hydrogen; quite certainly there will be no dust present initially. This is indeed a prime requisite of the present theory, for should any dust be present it will inevitably lead to the formation of molecules and these have the effect of lowering the temperature to well below that necessary for stars to form.

In the case of a supragalactic cloud of atomic hydrogen with an initial temperature in the lower range mentioned above, contraction will occur resulting in an increase in the radiation of energy by the hydrogen. As this happens, the temperature begins to fall until it reaches something like 10,000°K; it does not fall much below this as shrinkage continues. What happens next is that this vast cloud of gas begins to fragment into smaller systems, simply because owing to its large mass it is quite impossible for it to shrink as a whole. As Hoyle has shown, these smaller fragments will have masses about 10,000 million times that of the Sun, being comparable in this respect to the smaller galaxies such as M 33 in Triangulum and most of the others within the Local Group. Once these smaller fragments have formed, condensation continues up to a certain point; then further fragmentation occurs because, once again, the energy liberated by any further contraction of the mass as a whole would go into aerodynamic motions (those tending to produce a re-expansion)

207

rather than random thermal motions of the constituent atoms. And so the process is repeated until at the end of about four such fragmentations the individual masses are no longer transparent to radiation. In other words, they become sufficiently opaque to stop energy from radiating out into space. At this crucial point we find that the masses of these fragments lie between 0·4 and 1·6 times that of the Sun; that is to say, they are typical Population II stars.

The end product of this sequence is still a little different from the typical moderate-sized galaxies in that the system of stars produced is still a widely-scattered one rather like the halo that surrounds the galaxies. In time, however, there will not only be collisions between the stars but also a tendency to gravitate inward, and so a densely-packed region (the nucleus) is built up over a period of time. In addition to these random motions there will also be an overall rotation since the supragalactic cloud will possess a rotation itself, albeit quite a small one. As in all other condensing systems, however, the rate of rotation will increase as contraction occurs and it seems likely that this is the origin of the rotation of the galaxies. That galaxies do rotate is clear from the elliptical, rather than spherical, structure of the nuclei.

The foregoing process is therefore sufficient to explain how galaxies having masses similar to that of M 33 may be produced. What it does not explain, however, is the formation of the large spirals such as our own or M 31 in Andromeda (see Plate X). Nor does it indicate how the many dwarf galaxies originate. The difficulty is that only fragments having masses of approximately 10,000 million solar masses can be produced as the first stage in the fragmentation process. Larger and smaller fragments are not formed in this way. We must therefore turn now to the case where the initial temperature of the supragalactic cloud lies between 200,000 and 1,000,000°K to see whether this provides us with the explanation.

Here the situation is somewhat different from that we have just discussed in that the hydrogen does not begin to radiate until the cloud has shrunk to a very marked degree, but once the cloud has contracted to a diameter of around 2,500,000 parsecs radiation becomes considerable, so much so in fact that the temperature drops extremely rapidly again to around 10,000°K. The immediate result is fragmentation on a vast scale, but into smaller fragments than before, having a mass roughly 300 million times that of the Sun. This, then, explains the formation of the numerous dwarf galaxies, but what of the giants such as our own? The answer appears to lie in the fact that under these conditions agglomeration of a hundred or more of these small systems can occur, resulting in the formation of the huge galactic systems. Is there any observational evidence to support this idea of the compacting of many of these fragments in this way? In this respect, it is significant that the monster galaxies all appear to have attendant satellites in the form of the globular clusters, which are possibly the original small fragments still bound to them by gravity but not absorbed by them.

There still remains the question of the gas that did not condense into stars and which could amount to between 30 and 40 per cent of the total mass of the supragalactic cloud. This is almost certain to go into the formation of the inner disc due to collisions between one gas cloud and another, thereby dissipating the energy they have in the form of motion along the radius and at right angles to the galactic plane, although conserving their total angular momentum. Since there will be these violent disturbances within this gas, the possibility that magnetic fields may arise in it, eventually bringing into being the general magnetic fields associated with a typical galaxy, appears quite strong.

Once the stars of Population II have formed by the million the nuclear reactions taking place in their interiors, by which the heavier elements are built, will inevitably lead to supernova explosions, with resultant scattering of these elements throughout the mass of the galaxy. By this means, too, dust comes into being for the first time and, as we have seen, dust is one of the agents that will bring about a fall in temperature by means of the formation of molecules. Interstellar gas and dust clouds are thus formed within the spiral arms and are available for a second phase of star formation, these being the Population I stars of which our Sun is one.

Spectra of Galaxies

One of the characteristics of the galaxies that appears to vary fairly systematically with the evolutionary sequence mentioned earlier in this chapter is the spectral type of the nuclei. At the moment, a large number of galaxies have been examined in order to determine these spectral types. Certainly this implies an averaging of the individual spectra of millions of stars within these regions. Nevertheless, we find that the predominant type is dG3, very similar to that of the Sun, and in those spectra which have been recorded with a sufficiently high dispersion the dwarf characteristics are quite conspicuous. A relatively small number of galaxies are known whose nuclei show bright emission lines and other peculiar features and these – the Seyfert galaxies – will be discussed in the following chapter. In the meantime, we shall devote ourselves entirely to the 'normal' galaxies whose nuclear regions show predominantly absorption lines in their spectra. Here we find that the average nuclear spectrum varies quite systematically from about F8 or F9 in the open spirals to around G4 in the early elliptical galaxies. Certainly this range is small when we consider the whole range of stellar spectra but in spite of this it has been definitely established. There are, of course, certain discrete regions in both the irregular galaxies and the outer environs of the spirals where emission-type spectra are common, and undoubtedly these small regions correspond to the gaseous nebulae in our own Galaxy which are closely associated with hot bright stars, for example, the Orion nebula.

The presence of these hot supergiant and giant stars of spectral types O and B in the spiral arms of galaxies also appears to bear a relation to the evolutionary sequence. These were the stars which were first resolved by Ritchey and later by Hubble, and in the Sc galaxies they may be traced along the spiral arms to within a short distance of the nucleus. In the Sb galaxies, however, they appear only within the outer segments of the arms and they are totally absent from both the Sa spirals and the elliptical galaxies. In the case of the ellipticals, this can be explained on the basis of the relative absence of dust and gas, which are both necessary for the formation of these blue supergiants.

Colours of Galaxies

One further feature which, like the two just discussed, varies with the evolutionary scheme is the colours of the galaxies. Now the colour of a galaxy is a measure of the distribution of intensity along the underlying continuum and takes no account of the absorption lines which may be present. From the photoelectric measurements made by Stebbins and Whitford, using the 100-inch reflector at Mount Wilson, it has been shown that the elliptical galaxies have a colour excess of magnitude 0·3, this decreasing gradually until it vanishes at the point of the Sc spirals. Here the large number of blue supergiants that are present obviously more than makes up for the overall excess.

Sizes of Galaxies

Once the distances of the extragalactic nebulae were known, particularly the nearer ones for which the Cepheid method of distance measurement is available, their linear dimensions could be estimated directly from their angular dimensions. This was how Hubble first determined the sizes of the galaxies directly from photographs. The results he obtained showed that the extragalactic nebulae were far smaller than our own, for which Shapely had derived a diameter of some 300,000 light-years. By contrast, Hubble found values of 4,500 light-years for the typical elliptical galaxy and around 8,000 light-years for the spirals. This apparent discrepancy remained for more than twenty years, until two important facts were recognized. The first came with the discovery that interstellar space is not empty but contains obscuring matter which dims the stars and makes them appear more distant than they really are. The dimensions of our Galaxy are therefore reduced by about two-thirds, to a diameter of 100,000 light years. Second, sensitive photometric measurements of the galaxies have shown that the bright regions which appear on photographic plates are only part of their true size. Outer regions, which are too faint to be satisfactorily recorded photographically, extend their linear dimensions considerably. The original estimate of 9,000 light-years for M 31 in Andromeda, for example, has now been amended to approximately 180,000 light-years, making it somewhat larger than our own.

Rotation and Masses of Galaxies

The fact that our own Galaxy is rotating about a central hub has been demonstrated by the motions of the various types of stars. In the case of the other galaxies such a method is not feasible; nevertheless, methods do exist which not only demonstrate that such galaxies are rotating, but also enable us to measure the rates of rotation. Within the last two decades M 31 in Andromeda and M 33 in Triangulum, the two large spirals nearest to our own system, have been examined in detail by Babcock, Aller, Mayall and Wyse. Earlier we saw how many of the spirals show discrete regions which have emission spectra, these being the various gaseous nebulae, and by measurement of the Doppler shifts within the emission spectra it is possible to show that the nuclei of these galaxies rotate as if they were solid bodies while beyond these central regions the rate of rotation decreases fairly uniformly with increasing distance from the nucleus. From the rotational velocity gradient it is possible to work out the masses of the galaxies. That of M 33, for instance, is reckoned at about 2,000 million solar masses.

One point still remained to be cleared up, namely whether the spiral galaxies rotate in the same direction and whether this is such as to open out the arms still further or bind them more closely together. So far, this question has not been satisfactorily answered. Hubble has concluded from spectroscopic determinations that all galaxies rotate so that the spiral arms wind up, whereas Lindblad was of the opinion that the direction of rotation is opposite to this.

Intergalactic Matter

The presence of interstellar matter was firmly established some forty years ago but until 1950 there was no evidence of any kind to show that matter could exist between the galaxies. Rather they were regarded as utterly isolated systems lying in empty space. Since that time, however, irrefutable evidence has accumulated to show that this is far from being the case. Most of this evidence has been produced by the 200-inch Palomar telescope and the existence of diffuse material in intergalactic space is now generally recognized.

Let us begin with the environs of our own Galaxy. Long-exposure photographs have now shown that several faint blue stars are observable in the regions of the galactic poles, and from our knowledge of the intrinsic luminosities of such stars we can say with a fair degree of certainty that they are extremely distant, lying in intergalactic space beyond the periphery of the Galaxy. Whether they are actually stars which formed out in these regions or whether they are bodies that, at some time in the past have been ejected from within the Galaxy is still an open question.

The second line of evidence is based upon an observation that has already been discussed, namely that accurate photometric work has shown that the dimensions of the galaxies are much greater than would appear merely from photographic observation. The galaxy M 31 in Andromeda (Plate X), when examined by this method, is found to be sufficiently large to envelop the two satellite galaxies M 32 and NGC 205. Extending this argument still further, we may then envisage a situation where, within the very dense clusters of galaxies, the outermost environs of neighbouring galaxies are in actual contact.

Finally, several photographs taken with the Mount Palomar instrument show definite bridges of material between two galaxies, and in some cases even triple systems are known in which the individual members are thus linked. Now how can we explain the existence of these interconnecting bridges between galaxies? A clue to the answer may be found in the presence of single arms projecting from a point almost diametrically opposite that of the bridge, a situation very reminiscent of tidal effects, and this appears to be the most likely explanation. It also appears probable that the very peculiar radio sources which look like a pair of galaxies in collision may be two galaxies that are interacting in this way.

Galactic Red-Shifts

If we ignore the members of the Local Group which, as we have

212

seen, is a gravitationally bound system, we find that the absorption lines in the spectra of the galaxies all show a displacement towards the red end of the spectrum. If this is interpreted as a pure Doppler shift, it indicates that the galaxies are all receding from us with quite high velocities. This fact, coupled with the linear relation found between the velocity of recession and the distance, is of the greatest significance to cosmology and forms the basis of the idea of the expanding universe. It will be discussed in detail in Chapter 15.

11. The Seyfert Galaxies

The realm of the extragalactic nebulae has been extensively investigated only during the past three or four decades and even at the beginning of the present century the true nature of these objects was a matter of conjecture and dispute. On the one hand, there were those astronomers who considered them to be mere satellites of our own system, lying at distances of only 20,000 or 30,000 light-years, and on the other were the proponents of the island-universe theory, who believed that these nebulae lay at tremendous distances and were comparable in size to our own. Sir William Huggins had already demonstrated the existence of two classes of nebulae by means of the spectroscope—those that were purely gaseous and clearly lay within the boundaries of the Galaxy, and others whose spectroscopic features indicated that they were composed of many thousands or millions of stars.

With the advent of the 100-inch reflector at Mount Wilson, it became possible for astronomers to examine individual stars within the nearer galaxies and the discovery in them of the Cepheid variables and novae soon led to fairly accurate estimates of their distances. It then became evident almost at once that the latter view was correct. We now know that the extragalactic nebulae may be divided into two broad classes—the spirals, which may be seen full-face, edge-on, or at various inclinations to us, and the elliptical galaxies which show no evidence of spiral structure being smooth, featureless masses of stars, exhibiting a gradation from spherical configurations to elongated ellipses. A typical spiral galaxy such as M 31 in Andromeda (Plate X) contains some 100,000 million stars and throughout the whole volume of the observable universe there are many thousands of millions of galaxies.

In the present chapter we shall be discussing a small number – only a score or so – of extremely peculiar galaxies. The first of these was discovered as long ago as 1908, before the true nature of the galaxies was known. This was the spiral galaxy NGC 1068, found by Fath at Lick Observatory to show, in addition to absorption lines in its spectrum, several bright emission lines similar to those found earlier in such nebulae as that in Orion, which consists of a mass of glowing gas, Some nine years later a second, similar object,

214

NGC 5236, was discovered by Slipher who also showed that the emission lines in the spectrum of NGC 1068 are very broad, a characteristic that is typical of gases in turbulent motion.

It was not until 1943, however, that these objects came in for detailed study by Seyfert. Superficially, they resemble the normal spiral galaxies in their general appearance, but closer examination reveals that they have one very distinguishing feature apart from their curious spectra, namely a very small and intensely brilliant nucleus. It is this small, almost star-like nucleus that is the origin of the emission lines. The spiral arms themselves are indistinguishable from those of a normal spiral galaxy.

If we examine the spectrum of the nucleus of an ordinary spiral, for example M 31 in Andromeda, we find very few emission lines and those which may be present are all narrow, like those of the gaseous nebulae in our own Galaxy. The strange spectra, which we shall be discussing in detail later are, however, only one of the curious and puzzling features of the Seyfert galaxies. In a few of them the brightness of the nucleus varies over a period of a month or so and several are known that are powerful emitters of radio waves, the intensity of which varies violently over comparatively short periods. In addition, several of them radiate exceptionally strongly in the infra-red and all radiate in the ultra-violet.

We shall encounter these curious phenomena again when we come to discuss the quasars, and there is currently a widespread belief among astronomers that these two types of object may be closely related in some way. There are, however, certain well-defined differences between the quasars and the Seyfert galaxies. For instance, the emission lines in the quasars are all displaced far towards the red end of the spectrum, these large red-shifts indicating high velocities of recession and, if the red-shift-distance relation holds for them, they are also among the most distant of all known objects in the universe.

The Seyfert galaxies, on the other hand, show very little evidence of extreme red-shifts, those which have been measured being similar to those found in the relatively close spiral galaxies. Then, too, we know of no quasars which are surrounded by apparently normal spiral structures. In other words, apart from their curious nuclei, the Seyfert galaxies are identical with normal spirals.

The Continuous Spectrum

We already know that the way in which the intensity of the light from an astronomical object varies with wavelength in the

215

continuum gives us a very good idea of the temperature of the source and also of the character of the object. This variation is usually studied by comparing the intensity of the light that is transmitted by three special filters which allow only narrow bands of certain selected wavelengths to pass through, in the ultra-violet (U), the blue (B), and the yellow or photovisual (V). This three-colour photometry, as it is called, is much used for a variety of astronomical problems and if we employ it in the investigation of ordinary stars – or ordinary galaxies, for that matter, which are composed, in the main, of stars on the main sequence such as the Sun – we find that the photovisual intensity is the highest, followed by the blue and then the ultra-violet.

This is, of course, the sort of situation we would expect since most of the light from the stars comes through in the yellow and red of the spectrum, and it is only in the case of the very bright blue giants and supergiants that the B and U regions will predominate.

Suppose now that we use this method to investigate the Seyfert galaxies. What do we find? If the slit of the spectroscope is sufficiently wide to allow all of the light from the galaxy (including the outer, spiral regions) to enter the instrument, the intensities of light passing the U, B and V filters are very like those just mentioned. However, when we gradually narrow the slit to that only the light from the central regions is being examined, the blue and ultra-violet parts of the spectrum become dominant. The reason for this has presented astronomers with several puzzling problems which have still been only partially solved. It was at first thought that this excessive blueness of the continuous spectrum might be due to the presence of very hot, blue stars in the nuclei of these galaxies. This is not completely out of the question, since we might expect to find such young stars possibly embedded within huge gas clouds. However, from a study of such stars in our galaxy we know that they all show very distinctive absorption lines in their spectra, lines that are absent from the nuclear spectra of the Seyfert galaxies. It is only when we make allowance for the yellow radiation that is emitted both by cool stars and also by gaseous hydrogen, both of which are likely to exist within the nuclei of these galaxies, that we obtain a clue to the nature of these objects. When we do this we find that a blue continuum remains that is extremely like those produced by the quasars and also by certain supernova remnants, particularly the Crab nebula. Since their continuous spectrum follows a power law rather than Plank's law, Seyfert nuclei also show an infra-red excess relative to ordinary stars.

The Emission Spectrum

One of the nearest of the Seyfert galaxies, and therefore one of the easiest to study spectroscopically, is NGC 4151, which also has the advantage of possessing a very bright, star-like nucleus. Oke and Sargent have made a particular study of the emission spectrum of this galaxy, from which they have been able to deduce not only the volume and density of the material within the nucleus but also, by comparison with a more direct measurement of the volume of the nucleus, a great deal about its internal structure. At first sight this may appear surprising in view of the fact that, although one of the nearest of the Seyfert galaxies, NGC 4151, lies at a distance of some 10 million parsecs.

Let us see now how this information was obtained. The most prominent emission lines in the spectrum of this galaxy are those due to hydrogen, oxygen and nitrogen, with fainter lines of argon, sulphur and iron. Simply from the fact that these lines appear in emission, we know that the density of this material must be considerably lower than that found in stellar atmospheres. In other words, these emission lines are produced by gas, not in the form of stars, but as a diffuse medium between the stars. By measuring the relative strengths of the oxygen and nitrogen lines, Oke and Sargent were able to show that the temperature of this gas is of the order of 20,000°K, which is very like that prevailing in the Orion nebula. The planetary nebulae which are huge shells of glowing gas surrounding a very hot, central star – and which exist in fairly large numbers in our Galaxy – have similar temperatures. There is, therefore, nothing unique about the temperature conditions that exist in the Seyfert galaxies.

From the intensity of the hydrogen lines, it is possible to estimate both the mass and the volume of this gas by a fairly straightforward process. Hydrogen makes up almost the whole of it, the other elements constituting only about 1 per cent; furthermore, at this very high temperature the hydrogen will exist in the form of ionized atoms, that is, as equal numbers of free protons and electrons. The total numbers of these particles may therefore be determined from the intensity of the hydrogen emission lines, and the work of Oke and Sargent shows that the nucleus of NGC 4151 has a volume of 1,700 cubic parsecs and a mass of 20,000 solar masses. It was when the actual diameter of the nucleus was determined from observations with high angular resolution, however, that a surprising result emerged. By this method the volume was found to be not 1,700 but more than 60,000 cubic parsecs!

217

Such a large discrepancy cannot, of course, be explained away by experimental or theoretical error. Quite clearly there must be some other explanation. The suggestion by Oke and Sargent that the nucleus is not a coherent mass of gas but is made up of smaller clouds which are in violent motion has recently been verified by Anderson and Kraft, who have found that there are at least three fine absorption lines in the spectrum of this galactic nucleus, indicative of individual clouds of gas, each having a different velocity.

Earlier we mentioned that the emission lines are similar to those of the planetary nebulae. Here we find that these lines are broadened by the motion of the gas both towards and away from us – a Doppler shift broadening. If we extend this interpretation to the lines of the Seyfert galaxies we find certain puzzling observations. The velocities associated with the hydrogen lines are at least ten time greater than those of ionized oxygen – several thousand kilometres per second as compared with hundreds of kilometres per second. In addition, we also find that there are broad wings on either side of the hydrogen emission lines whereas none are found on the profiles of the ionized-oxygen lines.

There are so many similarities between the planetary nebulae and the Seyfert-galaxy nuclei as far as their spectra are concerned that many astronomers now accept that this line broadening is indeed a velocity effect. Accepting this, we must examine whether such velocities within these galaxies are random or ordered. If the motions are orderly they might be accounted for by rotational effects. If, on the other hand, they are chaotic, the most likely explanation is that violent explosive outbursts occur within the Seyfert galaxies.

It now seems probable that there is a small, dense core of hydrogen gas within the nuclei of these galaxies which is moving chaotically in all directions with velocities up to 5,000 kilometres per second, with the oxygen in a surrounding region of lower density. There are still, however, several difficulties associated with such a model. It is not easy to see how this gas, moving with such high velocities, can be held within this small core by gravitational forces, since this would require a mass of at least 20 million suns all within a sphere having a radius of only 0·05 light-year. We can overcome this problem in one of two ways: either the gas is not retained within the core and is being continually replenished in some way, or the velocities assumed so far are erroneous and there is some other effect coming into play within these galaxies which produces this abnormal broadening of the spectral lines.

Consider first of all that the values we obtain from the spectra indicate true velocities and the gas is not confined within the core by means of gravitational attraction. Then if there is no mechanism whereby it is being continually replaced, we must assume that the Seyfert galaxies represent merely a transient phase through which most, if not all, of the spiral nebulae pass during the course of their evolution. Recently, Andrillat and Souffrin have observed quite significant changes in the relative intensities of the hydrogen and ionized-oxygen lines in the spectrum of the central regions of one Seyfert galaxy, NGC 3516, and have proposed a model in which a rapid outburst of gas occurs within the nucleus of this galaxy.

This would tend to confirm the ephemeral nature of the Seyfert phenomenon but leads us to on a further grave difficulty for, although only about a score of Seyfert galaxies are known at the present time, they are found only among the giant spirals, of which about 800 are known with certainty. They are not, therefore, particularly un-common, numbering about 2 per cent of the total known. This makes it appear rather likely that each large spiral will undergo an explosive outburst of this kind about once a millennium. When we remember that out own Galaxy is a giant spiral of this kind, it appears surprising that we have no record of such an outburst having occurred at the galactic centre. The alternative is that the high velocities are not real and the broadening of the spectral lines is due to some other cause.

Photon Scattering

One of the possible mechanisms by which this may occur has recently been examined, namely the scatterings of photons by electrons. Now we often visualize light as composed of waves vibrating at various frequencies but, although in certain of its characteristics it does behave as though it has a wave form, in others it acts as if it were made up of discrete particles, these being known as photons. Now there are several lines of evidence which show that light possesses mass; for example, it can be bent by a sufficiently high gravitational field, as when the light from a star passes close to the Sun, and we may also show that light exerts a definite pressure, especially upon the heads of comets as they approach the Sun in their orbits around the solar system. The mass of a single photon is exceedingly small; nevertheless, photons may be scattered by atomic particles.

If we assume, therefore, that in the nuclei of the Seyfert galaxies there exists a gas in which protons and electrons are in equilibrium, both particles will have the same kinetic energy, given by the equation:

$$E = \tfrac{1}{2}Mv^2$$

219

where E is the thermal kinetic energy of the particle, M is its mass and v its velocity.

Now even if the velocity of the electrons is as high as 5,000 kilometres per second that of the protons, which give rise to the hydrogen emission lines in the spectrum, will be much lower, since the mass of a proton is much greater than that of an electron. The ratio of masses is approximately 1/1,800, hence the ratio of velocities will be approximately 42 (the square root of 1,800). The protons will therefore be travelling at about 120 kilometres per second, corresponding to a temperature of around 20,000°K, which is what Oke and Sargent found. Clearly such a model will depend upon the probability of a photon's being scattered by a particle as small as an electron, but calculation indicates that under the conditions prevailing in the nuclei of the Seyfert galaxies there are a sufficient number of electrons available to scatter each photon several times, thereby producing the wings found on either side of the hydrogen emission lines.

One other feature of the spectra of these galaxies which has been briefly mentioned is that they also show the lines of iron atoms which have been ionized several times (quite a high proportion of their outer electrons have been removed). We find similar lines in the spectrum of the Sun's corona and can be reasonably certain of the conditions under which they are produced. One of these conditions is that the temperature be around 6,000,000°K. So far the lines of ionized iron have been positively identified in only one Seyfert galaxy, NGC 4151. This is additional evidence for the model put forward by Oke and Sargent of an extremely hot, low-density gas in which there are cooler clouds having a somewhat higher density.

We must not, however, overlook the fact that the ionization of the iron atoms within the core may be brought about by quite a different mechanism from that of the very high temperature just mentioned. One possibility is the presence of powerful X-ray sources within the nucleus. It is certainly a well-established fact that X-rays have sufficient energy to knock the electrons out of their orbits around an atom. If this is the case, it may be possible to detect such sources by means of observations carried out by orbiting satellites equipped with sufficiently sensitive detectors.

Light Variations in Seyfert Galaxies

When we come to discuss the quasars we shall see that certain of them vary in brightness. Since the Seyfert galaxies show many similarities to the quasars, a programme of observation was undertaken to study possible light changes in these objects.

Photoelectric observations of NGC 4151 soon revealed such light variations, and shortly afterwards Kinman found similar changes in brightness in the quasar 3C 120, which has now been included among the Seyfert galaxies. As in the quasars, the cycle of these light variations can be as short as a month or so, indicating as we shall see later, that the diameter of the light source cannot be more than a few light-weeks.

Such observations serve to confirm the earlier findings that the nuclei of these peculiar galaxies are extremely small compared with the overall dimensions of the galaxies themselves.

Infra-Red Radiation from Seyfert Galaxies

A further characteristic of the quasars – that their peak radiation lies far in the infra-red region of the spectrum – stimulated a search for similar radiation from the Seyfert galaxies. Indeed, this was discovered before the light variability just mentioned. During an extensive search, Pacholczyk and Wisnewski found that the intensity of the radiation from NGC 1068 increases sharply as we move further into the infra-red and subsequent observations confirmed this for most of the Seyfert galaxies; the peak intensity lies in the far infra-red, in a region intermediate between that for a normal spiral galaxy, which is in the near infra-red, and that of a typical quasar, which although differing quite widely lie close to the longer wavelengths of the radio band (Fig. 41).

At first sight this intense infra-red radiation is somewhat surprising and several theories have been put forward to explain it. Of these, only one need concern us here since there are very grave difficulties associated with the others and it is generally accepted that they contribute little, if at all, towards the origin of this radiation.

When we examine the light from these galaxies, we find that it is polarized. Ordinary light, which is unpolarized, behaves like a wave that is vibrating equally in all directions at right angles to the direction of propagation. If the light vibrates more intensely in one direction than any other it is said to be polarized in that direction. Now one major cause of polarization of light is the presence of dust and certain of these galaxies are found to show dark lanes of dust. Another effect that dust particles have on light is a reddening, since the interstellar dust has a particle size of the right order to scatter the shorter wavelengths of blue light, but not those of red light. This reddening of starlight has been well established within our own Galaxy and it is therefore quite likely that most other

221

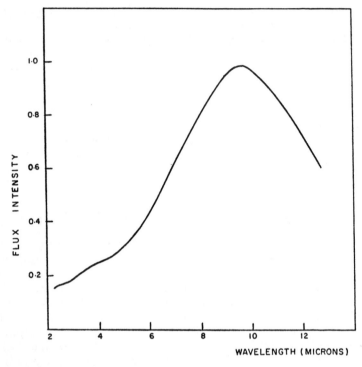

Fig. 41 The infra-red intensity curve of NGC 1068, the first Seyfert galaxy to be discovered.

galaxies, particularly the spirals, contain such dust clouds. Indeed, there are large numbers of galaxies in which this dust can be clearly seen as a broad, dark band along the galactic plane (plate xvi).

Having established the possible presence of dust within these peculiar galaxies, we must now see how it helps us to explain the intense radiation in the infra-red. This is not, as might be imagined, simply a scattering effect. Another mechanism comes into play, namely the absorption of energy by dust particles. Such a particle will readily absorb blue and ultra-violet light, resulting in a heating of the dust grain until an equilibrium is reached whereby it radiates as much energy as it receives. It remains now to determine the kind of radiation that is emitted by such particles. Estimates show that the effective temperature of these grains of dust will be in the region of 250 to 300°K, from just below freezing point to normal room temperature. The emitted radiation will therefore lie in the near and

222

far infra-red as observed in the nuclei of the Seyfert galaxies. Though this dust model is very attractive, it will face severe difficulties if alleged variations in the far infra-red on a time scale of months or a few years are confirmed.

Radio Emission from Seyfert Galaxies

The quasars, as we shall see in the next chapter, are powerful radio sources and, in view of the other similarities found between these two classes of object, we would expect to find evidence of radio emission from the Seyfert galaxies. While it is true that some of them are strong radio emitters, this is not so for them all. In some, for example, no radio emission has yet been detected, while others are very weak radio sources.

The general theory of radio emission suggested almost twenty years ago by Alfven and Herlofson postulates the existence of high-velocity electrons moving in a magnetic field. We have already seen that within the nuclei of the Seyfert galaxies there are electrons moving with velocities up to 5,000 kilometres per second. As Alfven and Herlofson have shown, once electrons are accelerated to velocities sufficiently high to take them into the region of cosmic radiation, they will emit radio waves by interaction with magnetic fields of the type and strength of those normally found within the galaxies.

Since it would be difficult to explain the optical continuum if the radio-quiet Seyfert galaxies and quasars possessed no high-energy electrons and magnetic fields, the most plausible explanation of the lack of radio emission is that these long wavelengths are absorbed by ionized hydrogen which is known to be present.

Relation of Seyfert Galaxies to Quasars

Throughout the whole of this chapter the similarities between the Seyfert galaxies and the quasars have been stressed. One further point now needs adding, to show that these similarities are more than merely superficial. At present we know very little of the process by which energy is being produced within the Seyfert galaxies or the quasars, although it appears to be some property associated with very large masses (of the order of 100,000 million Suns) compressed into an incredibly small volume.

One recent discovery, however, makes it seem fairly certain that the only distinguishing feature which separates these two objects into different classes is the amount of energy produced. A few years ago a small number of fairly bright, compact galaxies were found by Morgan, the so-called N-galaxies, which have energy outputs intermediate between the Seyfert galaxies and the quasars. This discovery provides strong evidence that here we have a continuous sequence of such objects, all having the common characteristic that they generate a vast amount of energy within a very small volume. At the moment it seems quite probable that astrophysicists will have to do a lot of rethinking of our knowledge of the way in which matter reacts under these incredible conditions before any understanding of the nature of these objects can be gained.

12. Quasars

The discussion of strong radio sources in Chapter 8 was almost entirely concerned with sources lying within the Galaxy – the clouds of hot hydrogen situated along the galactic plane, radio emission from the direction of the galactic centre, supernova remnants and the 21-centimetre line of neutral hydrogen. We must now extend the discussion to include a class of object which has provided astronomers and physicists with so many puzzling problems that we still know very little of its true nature and even the interpretation of the present observations is open to controversy. These objects are the quasi-stellar radio sources, often termed the quasars.

By 1961, radio surveys had revealed some 300 or so discrete radio sources. Among these were several for which no visual equivalents appeared on long-exposure photographs taken with the largest optical instruments; at the time it was recognised that much of the difficulty lay in the relative inaccuracy of the radio observations when it came to determining the positions of such sources. Unlike optical telescopes, radio telescopes have comparatively low resolution and it is far from easy to pinpoint a radio source with sufficient accuracy to enable it to be positively identified from a photograph of the region unless there is something obviously peculiar in the neighbourhood, such as a faint galaxy that may be a strong radio emitter or fine filaments of gas possibly associated with a supernova remnant.

The problem of determining the accurate positions of these quasars has been tackled from several directions. The first was by making use of lunar occultations of these sources. Naturally, this method is applicable only to those quasars which lie along the ecliptic but the positions of several such discrete radio sources have been determined by this method. It has the additional advantage, as Hazard has shown, that such observations provide detailed information concerning the diameter of the source in question, since we can use the limb of the Moon as a straight diffracting edge and obtain the diameter of the quasar provided that this is sufficiently small, from the diffraction patterns. The limiting accuracy at present is about one second of arc.

From observations carried out with the 210-foot steerable radio telescope at Parkes in South Australia during 1962 Hazard, Mackey

P

and Shimmins were able to determine the position of the very intense quasar 3C 273 extremely accurately from measurements made during three lunar occultations. Radio sources have been catalogued by the Cambridge astronomers and the designation of this particular source simply means that it is number 273 in the Third Cambridge Catalogue. Quasar 3C 273 which lies in the constellation Virgo, was observed at three different wavelengths, corresponding to frequencies of 136, 410 and 1,420 megacycles per second. From a study of the diffraction fringes, these workers were able to show that this quasar is a double source separated by 19·5 seconds of arc.

An accurate position having been found for the quasar, it was then possible to examine photographic plates taken of the region with the 200-inch Hale reflector for evidence of any object that might be identified with 3C 273. This was done by Schmidt, who found that the only objects close to the position given are a star of magnitude 12·6 and a faint jet of nebulosity which extends away from this star.

The Peculiar Spectra of the Quasars

As with most strange stellar or extragalactic objects, one of the first things to be done after a reasonably positive identification is to take its spectrum, since this generally proves of great value in determining the nature of such objects. The spectrum of 3C 273, however, turns out to be totally unlike anything previously known; it consists of a blue continuum upon which are superimposed several broad emission features whose positions could not be correlated with any known emission bands.

The explanation was eventually found in the presence of a very large red-shift. The postulated red-shift $\Delta\lambda/\lambda_0$ of 0·158 made it possible to identify six of the emission lines as those of the Balmer series of hydrogen and singly ionized magnesium and the forbidden line of doubly ionized oxygen. Even before these investigations another quasar, 3C 48, which was actually the first to have its position determined by interferometric methods at Cal Tech, was examined spectroscopically by Greenstein and Matthews. The spectrum of this source is similar to that of 3C 273 but at this stage the lines could not be identified because the familiar Balmer series was weaker than other lines which were later shown to be due to more highly excited ions at a very large red-shift. The clue to the identification of these lines was provided by the later work on 3C 273 and for coincidence of the observed lines with the Balmer series, an even larger red-shift of 0·368 must be postulated.

From this we can see that the spectra of the quasars, and several others have since been examined and found to have these abnormally large red-shifts – showing that these are very peculiar objects indeed. The spectrum of 3C 273 has been investigated by Oke into the infra-red region, confirming the results obtained in the visible and there now seems little doubt that these red-shifts are real.

The problem which faces us now is that of correlating these large red-shifts with apparently stellar objects. So far as we know, without putting forward completely new physical laws, there are two possible reasons for red-shifts of this magnitude. Either the quasars are stellar objects within, or close to, the Galaxy having gravitational fields of an exceptionally high order, resulting in a gravitational red-shift of the spectral lines, or they are extremely distant objects of an odd kind.

Let us examine these two possibilities in turn. We know that ultra-dense stars do exist, for example the pulsars to be discussed in the next chapter, which are believed to be neutron stars resulting from the collapse of a supernova remnant. If we postulate that the quasars are dense objects composed almost entirely of neutrons and other similar particles, we run into difficulties from the spectroscopic point of view. As we have seen, in the spectra of the quasars, we find both permitted and forbidden lines all having the same red-shifts and with widths that are only a small percentage of the wavelengh. On present theories of gravitational effects on spectral lines, this is not what we would predict. Either we would expect to find the different lines shifted by different amounts or the line widths would be greater than are actually observed.

The possibility that the quasars are very distant objects implies that the red-shift is cosmological in nature, that is, that it has to do with the relation of distance to velocity of recession; this must mean that these objects are among the most distant known. One particular radio source, 3C 295, appears to be receding at the fantastic velocity of 36 per cent of the speed of light. As far 3C 273 and 3C 48, their apparent velocities of recession are 47,400 and 110,200 kilometres per second respectively, putting their distances at 500 million and 1,100 million parsecs.

The Magnitude of the Quasars

Up to this point, we have been skirting around the major problem associated with the quasars, namely that of their visual and radio magnitudes. From the distance of 3C 273 and its apparent diameter as measured by the lunar-occultation method, we can calculate the diameter of the nuclear region that is emitting most of the radio energy and this has been estimated at about 1,000 parsecs.

3C 48, which is even more distant, has an absolute visual magnitude as great as —26·3, which may be compared with the brightest of the known giant elliptical galaxies, whose absolute visual magnitudes are —22·7. As far as the emission of energy in the radio spectrum is concerned, 3C 48 appears to be radiating about the same amount of energy as the most intense radio sources, for example, Cygnus A.

When we come to compare the apparent brightness of 3C 273, which is readily visible in even small telescopes, with that of galaxies at the same distance, it is not difficult to see why certain astronomers have sought an alternative explanation of the red-shift associated with the quasars. Although agreeing that the large red-shifts are indicative of high velocities of recession, they prefer to accept that the quasars are stars lying close to the Galaxy, thereby getting over the difficulty of explaining their extraordinary brilliance. It has been suggested that these tremendous velocities have been imparted to these objects by subsidiary explosions, somewhat of the nature of the Big Bang which resulted in the formation of the universe itself from the primordial 'super-atom'. Such an idea has to face several criticisms. If such gigantic explosions within the Galaxy did occur comparatively recently, we would expect such stars to be moving in random directions; at least some should be travelling towards us, exhibiting comparable blue-shifts in their spectra. In fact, no such blue-shifts have ever been observed. Secondly, we know of no mechanism by which ordinary stars would have such intense radio emission. While it is true that certain quasar-like objects are known which emit little or no radio energy, most of the quasars are powerful radio sources.

It is now accepted by most astronomers that the quasars are extremely remote objects and evidently the output of optical and radio radiation from a typical quasar must be on a truly stupendous scale, far exceeding anything else known in the universe. What kind of object can possibly emit such a tremendous flood of radiation and, equally important, are there any other means by which we can confirm the relatively small dimensions of the optical regions within the quasars?

Let us take the last part of this question first, since certain observations provide at least a partial answer. Over the past few years, a large number of quasars have been discovered, most of them characterized by strong radio emission and all by large red-shifts in their spectra. From a comprehensive study of several of these objects, some have been found which are not constant in brightness but show appreciable light fluctuations, with periods ranging from a few months to a year or more.

This is an extremely important observation since it tells us at once the maximum diameter of the light-emitting region within the quasar. Now we know from the spectra of the quasars that they are not, like the nuclei of ordinary galaxies, composed of large numbers of stars, even blue giants, which are among the most luminous known. The region producing the visible light appears to be coherent in the sense that it is approximately uniform in composition throughout its entire mass. Suppose, therefore, that we take the most extreme situation, namely, that the light variability originates in some part of the quasar farthest from us. Then clearly the light fluctuations must travel through the mass of the quasar with the velocity of light yielding a maximum diameter of between a few light-months and a light-year or so. The diameter derived in this way is considerably smaller than that calculated for 3C 48 by Greenstein and Matthews from direct measurements by the lunar-occultation method, which indicates that the optical and radio emission originate within a diameter somewhat less than 5,500 parsecs. Here, however, we must distinguish between the diameter of the optical continuum source (whose output varies) and that of the emission-line source (whose output is not known to vary and which can therefore be larger).

The unusually powerful optical radiation coming from so small an object would suggest that there are some basic physical processes arising within the quasars that cannot yet be explained. Some of the radiation appears to arise in electrons moving at relativistic speeds (approaching that of light) within magnetic fields, but other explanations have been given in the past. One of the most intriguing theories is that of gravitational infall. This is one of the few known mechanisms that can provide such a large outflow of energy.

Imagine that we have a large sphere of gas, consisting mainly of hydrogen, with a diameter of several hundred parsecs. If this gaseous sphere collapses very rapidly under the influence of gravity, the amount of gravitational energy released during the infall process will be exceptionally high. Now why should this be so? The answer lies in the fact that gravitational energy is proportional to the mass squared and inversely proportional to the distance between the surface of the sphere and its centre (the radius). Here, we assume that the mass remains virtually constant while the radius diminishes rapidly. The amount of gravitational energy released during the infall process will therefore rise rapidly in step with the decrease in the radius. If this energy, or a large part of it, were to go into heating the gas this might then account, in part, for the extremely high emission of visual light.

The Absorption Spectra of the Quasars

At this point we must consider one peculiarity of the spectra of some of the quasars that has led to intense speculation among astronomers. As we have already seen, the spectra consist mainly of broad emission features from the red-shift of which the distances have been determined on the assumption that the spectral lines obey the cosmological red-shift–distance relation. Several absorption lines have, however, also been discovered in the spectra of the quasars and here the situation is complicated by the fact that a typical quasar exhibits more than one red-shift for these absorption lines. If we assume that the red-shift is due merely to the Doppler effect, indicating a velocity of recession, then it should be the same for all absorption lines.

The first explanation of these differing red-shifts (which are found only for the absorption lines and not for the emission lines) was that it lay in some peculiar physics within the quasars themselves. This seemed to be in line with the extremely high radiation outflow, which can still not be fully explained on the basis of the ordinary laws of physics as we know them at present. However, this view has been challenged by some astronomers, since the emission lines do not show this peculiar effect and as an alternative it has been suggested that the absorption lines arise, not in the quasars themselves, but in intervening galaxies or clouds of gas through which the light from these distant objects has to pass before it reaches us. In the majority of cases these galaxies appear to be quite invisible and it is considered that they may represent dead galaxies, composed of dark stars which have long since passed into extinction, together with the remnants of gas and dust left over after star formation. On the basis of this idea, the absorption lines will partake of the velocities of recession of the individual galaxies, which may number up to half a dozen or so, each giving a different red-shift to the absorption lines in the quasar spectrum.

More recently, however, this idea has been criticized by G. R. and E. M. Burbidge, who have made a statistical analysis of almost 200 quasars for which sufficiently resolved spectra are available. They find that, when plotted on a histogram, the values of the red-shifts $\Delta\lambda/\lambda_0$ appear to fall around certain fairly well-defined maxima. The largest peak falls at 0·06, with a second at 1·95; very few indeed have values greater than 2·5. At one time it was believed that every quasar showed a red-shift of the absorption lines corresponding to the value of 1·95, but further study has now revealed several in which this particular red-shift is not observed. Rather than inferring that

230

the absorption lines originate within the intergalactic medium, the Burbidges believe that they arise in the quasars themselves; if this should indeed prove to be the case, then it is one more peculiar property that must be taken into account when any models of the quasars are put forward.

The Cutoff Beyond $\Delta\lambda/\lambda_0 = 2\cdot5$

The fact that few of the quasars have been found for which the value of $\Delta\lambda/\lambda_0$ is greater than 2·5 is one that clearly requires explanation. Although this large red-shift would indicate that such sources are extremely distant objects, we cannot simply dismiss the question by saying that such quasars are, at present, beyond the reach of the largest optical telescopes. A more plausible suggestion put forward very recently is that there exists some absorbing medium at these large distances which effectively obscures such quasars and it appears possible that this veiling medium is hydrogen, the remains of the primordial gas thrown out at a high velocity a few seconds after the Big Bang that marked the origin of the universe. There are, of course, many difficulties in detecting such distant clouds of hydrogen although Oke has found that one quasar, shown by Lynds to have a red-shift of 2·9, has a small amount of neutral hydrogen in front of it. Other such clouds may eventually be discovered by radio measurements of the 21-centimetre line which, like the absorption lines in the visible spectrum, should also show a large red-shift.

The Galactic Nucleus as a Quasar

So far we have been considering all quasars as very distant objects but for some time now it has been becoming increasingly obvious that in most, if not all, spiral galaxies the nuclei are the sites of very peculiar physical processes. Infra-red astronomers have now discovered that the centre of our own Galaxy is a very powerful emitter of infra-red radiation. Indeed, the intensity of this radiation is very like what we find in the nuclei of the Seyfert galaxies, described in the previous chapter. Whether the nuclei of all spiral galaxies show this strong infra-red emission is still a matter of speculation, but several teams of radio astronomers are at present working on this problem and results are expected in the very near future which ought to show whether or not this is so. At the moment, we can say that the available evidence does indicate a close relationship between normal and Seyfert galaxies and when the observations are extended throughout the observable spectrum it would seem that the relation between spiral and Seyfert galaxies is very similar to that between the latter and the quasars.

An estimate of the total number of quasars within the observable universe has been made by Sandage and from it Lynden-Bell has reckoned that the number of dead quasars in existence must be sufficiently high for one, at least, to be present within the Local Group of galaxies. The hypothesis put forward by Lynden-Bell to explain the link between the spirals, the Seyfert galaxies and the quasars, and also the tremendous infra-red emission from the central regions of the Galaxy is that there may indeed be a dead quasar at the galactic centre.

Here we must consider just what we mean by a dead quasar. From what we know of its characteristics, it appears that the lifetime of a typical quasar will be quite short compared with that of a spiral galaxy such as our own and furthermore it seems likely that, if there are quasars at the centres of spiral galaxies, their formation begins with the initial formation of the galaxy itself as it commences to condense out of the vast gas and dust cloud into individual stars. Owing, then, to its rapid evolution, the quasar will have spent all of its nuclear fuel by the time ordinary star formation is advanced, and because of its own tremendous mass it will suffer gravitational collapse. A stage will then be reached when the gravitational field of the quasar becomes so great that no light can possibly escape from it and, as far as optical astronomers are concerned, it will have ceased to exist. It will, however, still retain one powerful effect upon neighbouring matter owing to its intense gravitational field, and it is mainly for this reason that Lynden-Bell has suggested that such dead quasars will be found in the centres of large concentrations of mass, namely at the centres of the galaxies.

What then will be the nature of such a quasar? Basically, it is envisaged as a vast vortex of gas situated at the centre of the Galaxy, with matter being continuously drawn into it under the influence of gravity. Particles will be accelerated by changes in the magnetic fields within this whirlpool of gas and in the outer environs these will ultimately be changed into cosmic radiation, while the inner regions, owing to extremely frequent collisions, will emit intense infra-red radiation because the energy of the particles is converted into heat. Lynden-Bell has estimated that in order to account for the intensity of the infra-red radiation from the central regions of the Galaxy 0·001 solar mass of matter must fall into the quasar every century; correspondingly faster rates of infall would account for the nuclei of the Seyfert galaxies.

Investigation of the Solar Wind

Here we shall see how the quasars enable us to measure the density of ionized gas in interplanetary space. Apart from the debris left over from the formation of the solar system, dust, meteorites and asteroids, there are vast clouds of ionized atoms and free electrons, the density of which varies considerably, not only decreasing in a general way from the vicinity of the Sun out to the interstellar gas which lies beyond the confines of the solar system but also varying in a more discrete manner from one point to another. Some of this material undoubtedly arises in the Sun itself and forms part of an extended corona. The rest may come from beyond the solar system, being drawn toward the Sun by gravitational attraction, forming the splash corona mentioned earlier.

The free electrons detected in interplanetary space by Siedentopf and Behr have been shown to be spurious and we now know that the density of interplanetary electrons is much lower than they estimated. As far as the region close to the Sun is concerned, it is possible to measure the density of this 'gas' directly by photographic observation of the corona during a total eclipse. Similarly, a measure of the density out at the edge of the solar system can be found by direct observation at night from photographs taken of the sky in the direction away from the Sun.

The intermediate region, however, from the surface of the splash corona to about the orbit of the Earth, cannot be investigated by these methods and until recently no direct means existed for determining the density of these ionized particles. Now, however, we have the method of radio detection, used particularly by Hewish, Machin and Smith, which employs distant radio sources and not visual observation. Briefly, whenever radio waves pass through a region of space which contains free electrons, certain of their characteristics are changed so long as the intervening medium does not possess a uniform density. The changes that take place in the radio emission may be likened to the twinkling of starlight as it passes through the terrestrial atmosphere.

One essential condition for accurate measurement is that the radio source must have a very small angular diameter, and for this reason we cannot successfully use the more extended sources within our own Galaxy. The quasars, though, are ideal for this purpose and fortunately there are several which lie along the plane of the ecliptic and are nearly occulted by the Sun during its passage through these particular constellations. By comparing the radio emission from these quasars when they are far removed from the Sun with that we receive whenever the Sun passes close to their position along the ecliptic, we are able to examine the various alterations in intensity as they pass through different regions of the solar corona. In the next

233

chapter we shall see how this method was instrumental in the discovery of the family of peculiar objects known as the pulsars.

Finally, let us return to the question of the relation between the Seyfert galaxies, the Zwicky objects or N-type galaxies, and the quasars. The point was already made in the previous chapter that the type of energy produced by these three classes of object is identical, only the amount that is emitted being different in each case. This gradation of energy output throughout the sequence also appears to be in step with their distances. The discovery by Plagemann, Feldman and Gribbin that the distribution of the red-shifts among the quasars is random provides fairly convincing proof that these objects are, indeed, at cosmological distances and were formed at certain favoured intervals during the evolution of the universe. On the assumption that this view is correct, the quasars are by far the most distant of these three groups of objects. The Seyfert galaxies, although distant, are the nearest to us with the N-type galaxies occupying an intermediate position in space.

The problem is certainly a tantalizing one. In view of the grave difficulties so far encountered in attempts to define the nature of the energy production within these objects, it is possible that many of our present physical concepts of the behaviour of matter under these extreme conditions may have to be re-examined and that an entirely novel approach to physical laws is required.

13. Pulsars

We have seen how the quasars have proved extremely useful in investigations of the solar wind, vast clouds of charged particles hurled out into space by the Sun. The radio waves from extra-galactic radio sources are distorted as they pass through these 'plasma clouds', as they are called, but only if the source itself has very small angular dimensions. The ordinary radio galaxies such as the Seyfert galaxies already described – and even the radio sources within our own Galaxy – have dimensions that are much too large to cause this scintillation, which may be compared with the twinkling of starlight due to its passage through the atmosphere.

The discovery of the first pulsar by Hewish and his co-workers came in 1967 during an investigation of the quasars. By means of a radio telescope that had been specially built for the purpose of searching for quasars by the scintillation method (since this characteristic fluctuation of the radio waves is found only with objects of a very small angular diameter it serves as a simple method of differentiating quasars from other radio sources), and which operated at a wavelength of 3·7 metres, a large area of the sky was systematically searched. During the course of this survey, a large number of radio sources were found which were obviously quasars since they exhibited the scintillation phenomenon.

However, in August 1967, a very weak source was discovered that was flickering extremely rapidly and subsequent observations showed quite clearly, from its constant position in the sky, that this object was a celestial source and not the spurious result of a terrestrial radio signal picked up by the antenna. Since then, several pulsars have been discovered, notably at Cambridge and Jodrell Bank in England, Goldstone in California, Parkes and Molonglo in Australia and Arecibo in Puerto Rico.

It was perhaps inevitable that the strict regularity of the radio pulses should have led the early investigators to consider the possibility that here we had radio communication from some other extra-solar intelligence, and tests were carried out to determine if this could possibly be the case. In the first pulsar to be discovered, designated CP 1919, 15 pulses occur every 20 seconds, the pulse interval being 1·3370113 seconds. Very accurate timing, however, indicated that the interval between the pulses was decreasing slowly with time and this was soon shown to be a Doppler effect due to the fact that the Earth, in its orbit around the Sun, was approaching this particular source. If we make allowance for this Doppler effect, the pulse interval of 1·3370113 seconds remains constant to within one part in several million.

We can, of course, now determine whether, after the Doppler effect due to the motion of the Earth is eliminated, there is any residual Doppler shift, since if these pulses originate from some extra-solar planet, it too will be revolving around its parent sun, producing its own Doppler effect. All of the tests carried out gave negative results, however, indicating that no planetary motion other than our own was contributing to the observed shortening of the pulse interval.

Diameters of Pulsars

We saw earlier how the light variations in certain of the quasars have enabled astronomers to estimate their effective optical diameters; in the same way, measurement of the duration of the radio pulses from the pulsars provides an estimate of their dimensions, since radio waves travel at the velocity of light and the diameter of a pulsar cannot be larger than the time taken for such a pulse to travel across it. Now each pulse from a typical pulsar is not confined to a single wavelength but covers a range of frequencies and if we have a means of measuring the bandwidth of a pulse we can find its actual duration. For CP 1919, for example, the duration of a pulse is between 10 and 20 milliseconds and the maximum diameter is therefore around 5,000 kilometres – about half that of the Earth. The results obtained from the other pulsars are essentially similar, showing that all of these objects have dimensions like that of the Earth and are therefore far smaller than the Sun. Such a discovery immediately calls to mind the white-dwarf stars, which are well known. However, although comparatively large numbers of white dwarfs have been investigated for several decades, none are known to show this radio pulsation.

Galactic Position of Pulsars

As we have already seen, the quasars are generally considered to be extragalactic objects at very remote distances, possibly on the extreme edge of the observable universe. The very fact that the pulsars have such small diameters immediately implies that they are situated somewhere within the Galaxy, probably quite close to us. Until very recently, none of these objects had been identified with any optical source although many attempts were made to do this in the hope that perhaps they might show a similar regular fluctuation in the optical range of the spectrum.

In spite of this, it has been found possible to make reasonably accurate estimates of their distance. It is well known that the velocity of light in a perfect vacuum is slightly higher than in a dense medium, water for example, and the same thing applies to radio waves. In a vacuum, all radio waves, irrespective of their wavelength, travel at the same velocity. If, however, there are any ionized gases present, this no longer holds; the shorter the wavelength, the higher is its velocity.

Now interstellar space is not a perfect vacuum. This was first shown by Hartmann at the beginning of the century when he observed the K line of ionized calcium in the spectra of certain very hot, distant, B-type stars. In δ Orionis, which is an eclipsing binary, he found that, whereas all of the other spectral lines showed the characteristic oscillation due to the Doppler effect of the two stars as they revolved in their orbits around the common centre of gravity of the system, the K line remained stationary and was also very sharp and narrow, unlike the ill-defined hydrogen and helium lines. Further observations soon established that the intensity of these 'stationary' lines was proportional to the distance of these stars, confirming the assumption that they originate in a tenuous gas lying all the way between us and the star. What is more important in the present context is that, although the major constituent of these cool clouds is neutral hydrogen, a minor constituent is ionized hydrogen. This is only to be expected since hydrogen is the most abundant of the elements, but unfortunately the lines of ionized hydrogen do not appear in the spectrum and its presence can only be inferred indirectly.

The ionization of the hydrogen is brought about by ultra-violet radiation from very hot stars which strips away the solitary electron from the atom resulting in the presence of free electrons in interstellar space. The spreading out of a radio pulse, known as dispersion, is here due to a combination of large, cool regions with little ionized hydrogen and smaller, hot regions with a lot and, provided we know the density of electrons in space, we can calculate the distance such a pulse has travelled by timing the arrival of different wavelengths. The latter is comparatively easy to do. It is the difficulty of measuring the density of the ionized gas that introduces some uncertainty into the distance measurements by this method. Hewish and his colleagues have assumed a figure of 1,000 electrons in each cubic metre of space, yielding a distance to CP 1919 of 126 parsecs.

237

The nearest pulsar whose distance has so far been determined lies in the constellation Leo. This object, CP 0950, is only 30 parsecs away. Most pulsars, however, would appear to be of the order of several kiloparsecs distant.

Galactic Distribution of Pulsars

Quite clearly, all of the known pulsars are objects lying within the Galaxy and some are, cosmically speaking, very close to the Sun. When the positions of the first members of this class to be found are plotted, very little, if any, correlation is found with either the galactic plane or the galactic halo. Indeed, the pulsars appeared to be scattered fairly randomly throughout the Galaxy, some lying close to the Milky Way (CP 0328, PSR 1749), and others in the region of the galactic poles (CP 1133).

More recently, however, several new pulsars have been discovered at Molonglo Radio Observatory and these show a distinct tendency to cluster in groups of two or three along the galactic plane. The behaviour of these newly-discovered pulsars is quite noticeable and there is now strong evidence to indicate that they lie close to, or actually in, the spiral arms of the Galaxy. This tendency to cluster in small groups may possibly be associated with the explosions of supernovae and it is to this subject that we must now turn.

Nature of Pulsars

The fact that the pulsars are all extremely small objects, of planetary size, suggests that we are dealing either with white dwarfs or with neutron stars. Until the discovery of the pulsars, the latter objects had been described only as theoretical models and had not actually been observed. At this point, it is appropriate to consider the properties of white dwarfs and neutron stars and how they fit into the scheme of stellar evolution outlined in Chapter 4.

The white dwarfs have long been recognized as abnormal stars which do not obey the mass-luminosity relation first put forward by Eddington, being far too massive for their luminosities. Since their masses are generally between 0·6 and 1·5 solar masses, all of which is compressed into a volume of planetary size, their densities are extremely high – about 100,000 times that of the densest known metal. How is it possible to explain such high densities? The answer lies in the fact, as shown by Fowler and later fully investigated by Fermi, that at these very high pressures the electrons in the atoms are forced out of their normal orbits, the orbital energy they would normally possess being translated into random motion. We can therefore envisage a 'sea' of electrons in which the other nuclear particles are embedded. This degenerate matter, as it is called, has the unique property that the larger its mass, the smaller is the volume it occupies. The critical pressure above which normal matter collapses into degenerate matter is of the order of 10 million atmospheres. As we have seen, the white-dwarf stage represents, for most stars, the final evolutionary phase before total extinction.

Suppose, however, that conditions are such that a somewhat more massive body collapses in a similar manner. Theoretically, the internal pressure will be higher than in the white-dwarf star. When this happens, the electrons will be forced into the nucleus, combining with the protons to form neutrons. We shall then have a form of degenerate matter composed of a 'sea' of neutrons rather than free electrons.

Such an object will be a neutron star and it is now believed that these may be formed by supernova explosions. The first important clue to suggest that pulsars may be the hypothetical neutron stars came with the discovery of two pulsars that are closely associated with known supernova remnants; these are the Crab pulsar, which lies close to the centre of the Crab nebula in Taurus and the Vela pulsar, which is identified with Vela X, a suspected supernova shell. It is significant that these two pulsars have the shortest periods so far known, that of the Vela pulsar being 89 milliseconds and that of the Crab pulsar only 33 milliseconds.

There are several theoretical reasons for believing that the pulsars are neutron stars rather than white dwarfs. To begin with, if a pulsar such as CP 0950 or CP 1133, which are estimated to be only 30 and 49 parsecs distant, were typical white dwarfs, we would expect to see them with the 200-inch reflector as very faint optical stars. The fact that, so far, they have not been observed visually or photographically indicates that they are not white dwarfs, although it does not strictly prove it.

We also have to explain the emission of the radio pulses from these objects and at present there are two widely different theories in vogue. Since there are points in favour of both, we will here discuss these in some detail.

Pulsation Theory of Pulsars

We know that the Sun emits radio waves, the reason for this being that sonic waves emanating from the photosphere travel upward through the solar atmosphere, increasing fairly rapidly in velocity as the density of the atmosphere diminishes, until a shock wave is produced. The effect of such shock waves will be to accelerate electrons within the outer atmospheric levels to high speeds, and

239

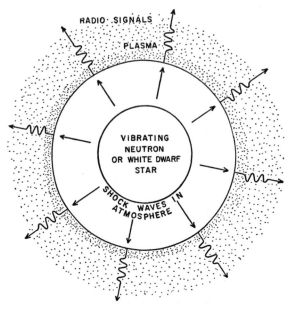

Fig. 42 The emission of radio waves from a pulsar according to the pulsation theory. Shock waves travelling through the star's atmosphere are transformed into radio waves by interaction with the surrounding plasma.

these electrons will then generate radio waves as they move through the ionized streams of gases surrounding the Sun. We have already seen how the solar wind is composed of clouds of ionized particles ejected from the surface of the Sun into interplanetary space. According to the pulsation theory of the pulsars, a similar effect is produced by an oscillation of the entire atmosphere, the frequency of these oscillations being the same as that of the radio pulses we receive (Fig. 42).

Now to what extent does our knowledge of the properties of white dwarfs and neutron stars support this picture? Is it possible for such objects to oscillate as rapidly as will be necessary for the very high frequency of the radio emission? Before the discovery of the Crab and Vela pulsars a grave difficulty arose in that, according to the calculations made, the white dwarfs cannot vibrate sufficiently rapidly for the observed frequencies of the radio pulses and the neutron stars appear to vibrate too quickly. When it was shown that two pulsars, at least, have the extremely short periods of only

33 and 89 milliseconds, this difficulty was almost completely removed as far as neutron stars are concerned. There is now little doubt that these objects can vibrate with periods as short as this, whereas if a white dwarf were collapsed sufficiently to vibrate with a period of less than a few seconds it would collapse utterly under the influence of gravity. The only way out of this dilemma is to assume that the atmospheric levels vibrate at the rate necessary for the production of the radio waves, this rate being coupled in some manner to a somewhat slower rate of vibration of the interior of the star. On the whole, however, if we accept the pulsation idea, the neutron-star theory appears to fit the observed facts far better, particularly in the case of those pulsars having very short periods.

Rotation Theory of Pulsars

The idea that an active spot on the surface of a rapidly rotating white-dwarf star might send out a beam of radio waves, analogous to the sweeping light beam from a lighthouse, has been advanced and developed by Ostriker. Such a theory may account for those pulsars that have relatively long periods, of the order of a second or more, but if we attempt to apply it to pulsars such as that within the Crab nebula we immediately run into theoretical difficulties. Such a star would require to rotate about thirty times every second and for a body of the size and constitution of a white dwarf this would almost certainly result in complete disruption or a very uneven rate of rotation and the latter could not possibly account for the regularity of the emitted radio pulses. A neutron star, on the other hand, has a diameter of between 10 and 100 kilometres and can spin at this prodigious rate without disruption. Gold has proposed a model of a rapidly-spinning neutron star surrounded by an ionized gas, or plasma, that is held very tightly to the star itself by an exceptionally strong magnetic field. Since highly ionized matter cannot normally cross magnetic field lines owing to its high electrical conductivity, this means that the plasma rotates with the star and, being of a much larger diameter than the star itself, the speed at the periphery of the plasma is very close to that of light and it is here that the electrons, travelling at this extremely high velocity, cause the emission of radio waves (Fig. 43).

When we look further into the details of such a model we find several features that are in striking accord with recent observations. On the basis of this model Gold predicted that, since neutron stars are believed to be formed only during a supernova explosion, they will be associated with the positions of known supernovae. This has been borne out by the discovery of the Crab and Vela pulsars. Also,

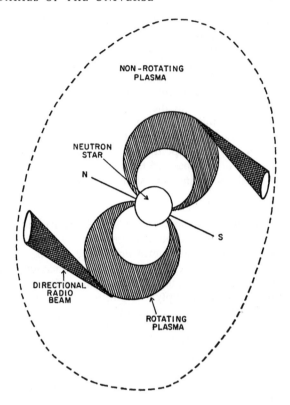

Fig. 43 One of the lighthouse theories of pulsars; radio emission is due to electrons moving with relativistic velocities at the periphery of the rotating plasma clouds.

since it is theoretically possible that a supernova explosion may produce more than one neutron star, owing to the disruption of the supernova, this could account for the discovery of the small clusters of two or three pulsars found by the workers at the Molonglo Radio Observatory. A further feature of the pulsars is that the period and the duration of the pulse show a definite correlation, and finally there should be a detectable slowing down of the pulse rate. This last prediction, also made by Gold, is worthy of some additional comment.

When a neutron star is newly formed during the last stages of a supernova explosion, we can estimate its rate of revolution from the angular velocities associated with a rapidly contracting star and it is not inconceivable that at first the neutron star will rotate in about 1 millisecond. This rotation will not be maintained for long, however, since energy will be lost quite quickly on account of instabilities and the ejection of gas into the surrounding plasma. The result will be a slowing down of the rate of revolution which will continue as energy is lost, either by the loss of matter or by radio emission. Accurate timing of the pulses from the Crab pulsar by Gold has shown that such a change in the period of the radio emission is actually occurring. The change is quite small, amounting to only about one part in 2,400 each year, but since the pulses are so rapid and regular this is well within the limits of detection. Other pulsars, particularly CP 1919, CP 0834, CP 0950, and CP 1133, have also been shown to exhibit the same effect by Davies, Hunt and Smith.

Generation of Cosmic Rays by Pulsars

As may be appreciated, the energy associated with a typical pulsar, particularly during the early stages after its formation, is extremely high, so high in fact that much of it goes into the production of cosmic rays. At the moment it appears highly probable that the greater portion of the cosmic-ray flux within the Galaxy comes from rapidly rotating neutron stars.

Light Radiation from Pulsars

One striking feature of the pulsars is that, although it has proved relatively simple to detect the radio emission from these objects ever since the discovery of the first pulsar indicated the range in which these pulses lie in the electromagnetic spectrum, the detection of light from these stars has proved to be extremely difficult. This, we now know, was largely due to the manner in which the search was carried out. Very intensive surveys of the regions of the known pulsars have been made with the largest instruments, without any initial success. Only recently, when it was recognized that any emission from these objects in the visible region might be in the form of short pulses similar to the radio waves, have flashes been detected from the Crab pulsar.

The method used consists in taking a series of photographs in very rapid succession by means of a highly sensitive film and highly sophisticated techniques to enhance the amount of radiation falling upon the film. Undoubtedly, with further improvements in this technique, optical flashes from other pulsars will be detected.

243

14. Extraterrestrial Life

We must now consider one of the most interesting and intriguing problems posed by astronomy, a problem to which the study of certain close binary systems and more recently the chemistry of meteorites and the pictures of the Martian surface obtained by Mariner 7 has now provided at least a provisional and encouraging answer. For several thousand years the various religions and superstitions of the world taught that Man is alone in the universe, that only upon this one planet, which is but one of countless millions within the universe, has thinking life evolved. Even as recently as the beginning of the present century scarcely any evidence existed in support of the possibility that life might have come into being elsewhere than on Earth. Even with the understanding that the Sun is a very ordinary star, the idea of extraterrestrial life was disregarded by most astronomers, such a belief being influenced a great deal by the then current theories of the evolution of our planetary system, which suggested that is was more a freak of nature than a fairly common occurrence.

Over the past few decades, the general concensus of opinion among astronomers has swung the other way and it is now considered much more likely that planetary systems are quite commonplace among certain types of star. Before we extend our survey into interstellar realms, however, let us first examine our own solar system for any evidence there may be of other life forms. Unless we venture into the field of pure speculation we must, of course, impose certain conditions. All life as we know it is based upon the element carbon, which has certain unique properties; chiefly among these is the ability to combine with itself to form extremely complex structures, a property shared by no other chemical element with the possible exception of silicon (even here those compounds which are sufficiently stable and complex have silicon atoms linked through an intermediate oxygen atom and not to themselves). A further characteristic of life is the need for respiration (here we exclude the viruses, which seem to occupy a halfway point between living and inanimate matter), and respiration requires the presence of an atmosphere of some kind. We must also impose certain limits upon the environmental temperature and finally, as we shall see later, the presence of liquid water is probably a necessity.

244

Of the planets of the solar system, Mercury and Pluto can be ruled out at once, the former because of its close proximity to the Sun and lack of any appreciable atmosphere, the latter owing to its extremely low temperature, close to absolute zero, which will almost certainly have frozen any atmosphere the planet may once have had on to the surface.

The environments of Venus and Mars, the most likely candidates for supporting life, particularly intelligent life as we know it, are extreme to say the least, although they may yet be found to be compatible with the existence of some low form of life. We know very little of the surface conditions on Venus. The temperature near the surface is somewhere in the region of 300°C and any oxygen or water vapour that may be present in the Cytherean atmosphere is certainly very low compared with the very high concentration of carbon dioxide. The intensity of the ultra-violet radiation that penetrates the atmosphere is also far in excess of that reaching the surface of the Earth and, as we shall see, this type of radiation is very effective in destroying life. Our knowledge of the surface conditions on Venus is still extremely meagre. The gross features of the planet are strikingly Earth-like. The mass, density, volume and escape velocity are all similar to that of our own planet. The major puzzle is that of the Cytherean day, the time taken for Venus to rotate upon its axis. Various conflicting estimates have been made over the last three centuries ranging from a value of 23 hours 21 minutes given by Cassini in 1666, to 224 days 16 hours 48 minutes suggested by Schiaparelli in 1890, the latter being the same as the sidereal period of Venus, indicating that it kept one face turned perpetually to the Sun.

All of these early estimates were obtained by the observation of bright or dark shadings on the planet and, since these features are merely transient atmospheric phenomena, little real significance can be attached to them. Recent radar methods and evidence obtained from both American and Russian rockets which have made close approaches to the planet show that the Cytherean day is certainly much less than the 224 days put forward by Schiaparelli. At present we can only say that, although conditions on the planet are extreme and the dense atmosphere is not one which would support animal life as we know it, we cannot completely rule out the existence of some form of life on Venus. The final proof will come only when Man sets foot on the planet.

Our ideas of the conditions existing on the surface of Mars have altered dramatically since the very recent publication of the excellent photographs taken by Mariners 6 and 7, one of which is shown in

245

Plate V. We have already discussed these in Chapter 2 and here we will only concern ourselves with those features which have any bearing on the possibility that life exists on this planet. The mean temperature of Mars is, as we would expect from its greater distance from the Sun – 141 million miles as compared with 93 million miles for Earth – appreciably lower than on our own planet, being about −25°C compared with +15°C. During the Martian night, however, the temperature frequently falls as low as −40°C at the equator, while in the polar regions it can be as low as −100°C. As in the case of Venus, the intensity of the ultra-violet radiation at the surface is quite high although for Mars this is due to the thinness of the atmosphere and not to its proximity to the Sun. Quite clearly, from these figures alone, any astronauts who land there some time in the future will require protection from such alien environment.

The Mariner fly-bys of the planet have also revealed the complete absence in the Martian atmosphere of nitrogen, a gas which, although it makes no direct contribution to human respiration, nevertheless plays a very important role in the complex series of reactions that lead to life coming into being. It is, however, significant that observations made close to the southern Martian pole have indicated the presence there of methane and ammonia, two gases that are believed to have played an important part in the early stages of life on Earth when, as we shall see, the atmosphere was quite different to that we now know. In addition to these two very important gases, a small amount of water vapour has also been detected on Mars, although whether the polar caps are made up of ice or solid carbon dioxide is a point which cannot yet be answered with any degree of finality.

Certainly, this latest information we have concerning Mars has exploded many of the old theories, and the idea that Mars is a much older version of Earth, where life came into being many millions of years ago and has since died out, must be abandoned. It would be completely wrong, however, to conclude that both Mars and Venus are completely sterile, as has been proved in the case of the Moon, although this is a point of view held by some astronomers. We know, for example, that there are certain terrestrial organisms which can withstand ultra-violet radiation that would be lethal to higher life-forms, which could also survive and adapt to the conditions existing on both of these planets. Some astronomers also believe that we may find evidence of fossil life on other planets of our solar system, evidence that higher life-forms once existed beyond the Earth in past ages.

The existence of life on the planets of other stars in the Galaxy is now a point worthy of serious consideration for clearly, if we are to find life comparable with our own, it is here that we must seek and not among the solar planets. However, before we discuss this much larger question of extra-solar planets, we must first look at the origin of life on our own planet and then at the evidence that has recently been obtained for the presence of certain organic compounds in the centres of carbonaceous meteorites.

The Origin of Life on Earth

Two thousand years or so ago, Aristotle put forward the concept of spontaneous generation of life from completely inorganic material, the general idea being that life began abruptly from inert matter. The Aristotelian theory of the origin of life remained unchallenged throughout the Middle Ages, with only minor modifications. Not until the invention of the microscope and the early researches of van Leeuwenhoek were any serious doubts cast upon this hypothesis. Once the existence of bacteria was demonstrated, it was possible to show that many of the so-called changes taking place could be attributed to these extremely tiny organisms. The final blow to the Aristotelian idea came with Pasteur's work, which proved conclusively that no such phenomena occurred under completely sterile conditions.

There then followed several decades during which no acceptable alternative was forthcoming, although reports continued to be published claiming to offer proof of living creatures found in objects of meteoritic origin. All were, however, shown to be based either upon erroneous observations or upon contamination of these meteorites after their fall to Earth.

Gradually the old theory was resurrected, but in a vastly modified form. No longer was it considered that life came into being abruptly, rather that the process was an extremely slow one, taking place over several million years, developing from the molecules that formed the early planet.

Here we must recognize one very important point. The atmosphere of the primal Earth was far different from that which it has at present. A simple hypothesis which was put forward as recently as 1928 by Haldane and Oparin has enabled astronomers, chemists and biologists to formulate a plausible theory of the way in which life on Earth evolved by means of a tremendous and complex mechanism of selection and chemical transformations; several of the key steps in this chain of molecular events are now susceptible to experimental observation in the laboratory.

The atmosphere surrounding the planet at the beginning of its

247

existence was a reducing one as opposed to the oxidizing atmosphere which it now possesses. The gases of which it was composed were similar to those we now find in the giant planets – Jupiter, Saturn, Uranus and Neptune. By far the greater portion of it was made up of hydrogen, methane, ammonia, nitrogen and water with carbon monoxide and carbon dioxide present in somewhat smaller quantities. In all probability, the surface temperature of the Earth was then far higher than it is now and, over a period of time, this primeval atmosphere was lost into space. Following this atmospheric loss, oxygen was gradually produced, chiefly by dissociation of water by the action of light, a process known as photolysis. Later, once chlorophyll – the green matter of plants – was synthesized, there was a further liberation of oxygen from carbon dioxide.

We can thus visualize the oceans of Earth in those early days as a kind of 'soup' containing dissolved gases and simple molecules which were still a long way removed from anything we can conceivably regard as life. We must not forget, however, that other conditions on the planet were also far different from those we know today. Temperatures were generally higher, storms were more frequent and intense and the incident ultra-violet radiation from the Sun was at a much higher level. All of these, together, with the intensity of the incoming cosmic radiation, were sufficient to bring about the formation of reactive molecules which, particularly in solution, linked up to form simple amino-acids which are the building blocks of living things.

Until 1951, all of this was little more than a speculative theory. Since then, however, several scientists, in particular Miller, have succeeded in simulating the conditions which could have existed on Earth all those millions of years ago and have demonstrated experimentally that these compounds are indeed formed under the influence of electricity or ultra-violet rays. The step from a complex 'soup' of such molecules to even the simplest of living cells is a very big one indeed and obviously we still cannot say just when the spark of life was struck. All we can say with some degree of certainty is that it must have happened, over several millennia, at more than one point on the Earth. In other words, life did not begin in one particular spot in the ocean and spread outward, finally engulfing most of the planet.

As to the period in the history of the Earth when life actually came into existence, the further back in time we go the more difficult it is for scientists to differentiate between those chemical compounds which are not necessarily formed by living organisms and those which are truly indicative of life itself.

248

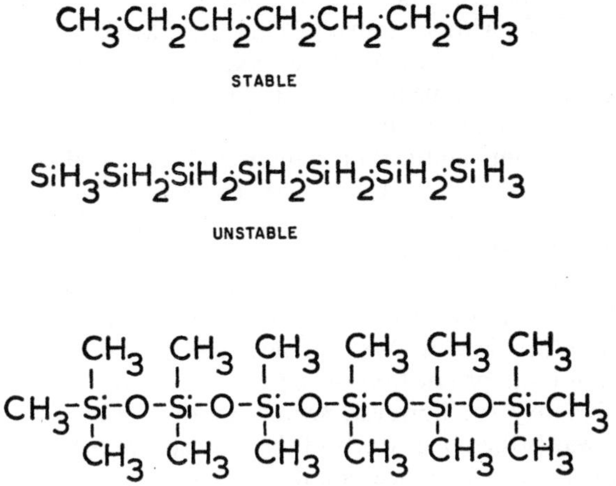

$$CH_3 \cdot CH_2 \cdot CH_2 \cdot CH_2 \cdot CH_2 \cdot CH_2 \cdot CH_3$$

STABLE

$$SiH_3 \cdot SiH_2 \cdot SiH_2 \cdot SiH_2 \cdot SiH_2 \cdot SiH_2 \cdot SiH_3$$

UNSTABLE

$$\begin{array}{ccccccc} & CH_3 & CH_3 & CH_3 & CH_3 & CH_3 & CH_3 \\ & | & | & | & | & | & | \\ CH_3- & Si-O-Si-O-Si-O-Si-O-Si-O-Si-CH_3 \\ & | & | & | & | & | & | \\ & CH_3 & CH_3 & CH_3 & CH_3 & CH_3 & CH_3 \end{array}$$

STABLE

Fig. 44 The relative stability of carbon and silicon compounds. Only when the silicon atoms are linked through an oxygen atom are complex silicon compounds stable.

The Characteristics of Life

A lump of rock or a grain of sand is clearly not a living thing. A single bacterium, a plant or an animal equally clearly is. What, then, is the dividing line between these two groups of matter? To begin with, we find that organic materials are, almost exclusively, carbon compounds. Now, as we have mentioned earlier, the element carbon possesses the unique ability of combining with itself to form extremely complex structures unlike those of any other element. The closest approach to carbon is the element silicon but, while it shows a somewhat similar behaviour, it does so by linking through an oxygen atom and not to itself (Fig. 44).

The two main characteristics of living things, however, are first that they exist in what is known as the colloidal state, being neither solid, liquid nor gaseous but a suspension of very small particles in a water medium (only rarely do we find non-living matter in the colloidal state), and second that they have the ability to assimilate food and convert it into the energy necessary for their growth, development and reproduction.

At some time during the early history of the Earth, therefore, the intense ultra-violet light from the Sun, coupled with the tremendous electrical discharges within the atmosphere, brought about the syn-

249

thesis of large organic molecules, extremely complex but still only on the verge of life, without actually being alive. Much later, within the sterile 'soup' of the oceans, these molecules formed into colloidal gels surrounded by a membrane, very like those of living cells. But it was not until the appearance of chlorophyll – the green matter we now find in plants and certain minute organisms – that a very real change in the environment occurred. The reason for this change is well known. Chlorophyll has the ability to absorb light of a much longer wavelength than that of ultra-violet radiation and will also convert carbon dioxide, which was then quite plentiful in the atmosphere, into oxygen which could be utilized to synthesize further large molecules. We can now see how one of the major processes for the production of oxygen came into being.

This free oxygen, however, served another important function. Rising towards the upper levels of the atmosphere (since oxygen is a lighter gas than carbon dioxide), it was acted upon by the incoming ultra-violet radiation to form a layer of ozone, which effectively shields the atmosphere below it from the ultra-violet rays that would be quite lethal to living things on the surface of the Earth.

One of the great difficulties facing scientists in their search for the earliest forms of life is that these creatures, as yet, possessed no bony structures like those of the animals which came later. As a result, although we have a very good idea of the types of organism which may have existed, the search for their actual remains has met with only meagre success. It is understandable, then, that relatively few geologists have concerned themselves with the beginnings of life on this planet, leaving the search to the biochemists and the astrophysicists. The presence of algae has been confirmed in rocks dating back to more than 2,000 million years ago but, if we wish to search back in time still further than this, we must have recourse to some other method. Fortunately, within the last two decades, tremendous advances have been made in analytical techniques which have made it possible to trace back the origins of life to more than 2,700 million years ago.

The method depends upon the isolation and identification of certain compounds known as long-chain hydrocarbons since they contain only carbon and hydrogen in the molecule; these compounds, it is argued, can only have been formed by living organisms. Although there is still some controversy over the biogenic nature of these compounds, other lines of evidence have, as we shall see later, supported the idea that this is indeed so.

Evidence for Extraterrestrial Life

We must now consider the evidence which suggests that life may exist, or may have existed, elsewhere in the solar system. Two lines of investigation, one indirect, the other of a more direct nature, have led many scientists to believe that the Earth is not the only body in the solar system on which life originated.

Theoretically we may argue that, once in existence, life will adapt itself over a sufficiently long period of time to its environment. We must therefore be very careful not to be too dogmatic in asserting that life cannot exist merely because the observed environment is different from that which we know on Earth. The long-held belief that the Moon is sterile has been proved following the recent landing of men there and the subsequent examination of the lunar rocks brought back to Earth. A great deal of information, too, has been received from the unmanned probes sent to Venus and Mars, so that now we have a good idea of the conditions that exist there. Such information has been used to simulate these conditions quite accurately in the laboratory in order to determine how certain well-known organisms would react if exposed to these environments.

Naturally, until men actually land upon these planets, some uncertainties will remain but at the moment there seems little doubt that some low life-forms could live and adapt to conditions such as those on Mars. It is extremely important, therefore, that any probes designed to make a soft landing upon either of these planets should be completely sterile; otherwise, when manned landings are made, it will be impossible to determine whether any microscopic life that may be discovered is truly indigenous to the planet or is merely a contaminant, taken there previously from Earth.

Chemical Composition of Meteorites

Apart from the recently obtained lunar rock samples, the only bodies of extraterrestrial origin that we can examine physically in the laboratory are the meteorites. The term meteorite is reserved for those bodies which enter the Earth's atmosphere from outer space and are sufficiently large to survive the fall to the surface without burning up completely. Broadly speaking, meteorites may be divided into two main classes – the siderites or ferrous meteorites, which are composed mainly of iron with a trace of nickel, and the stony meteorites. In the latter group we find a special class, known as chondrites, which have a peculiar granular structure; of these, a small number, termed the carbonaceous meteorites, have a very high organic content, often amounting to several per cent by weight.

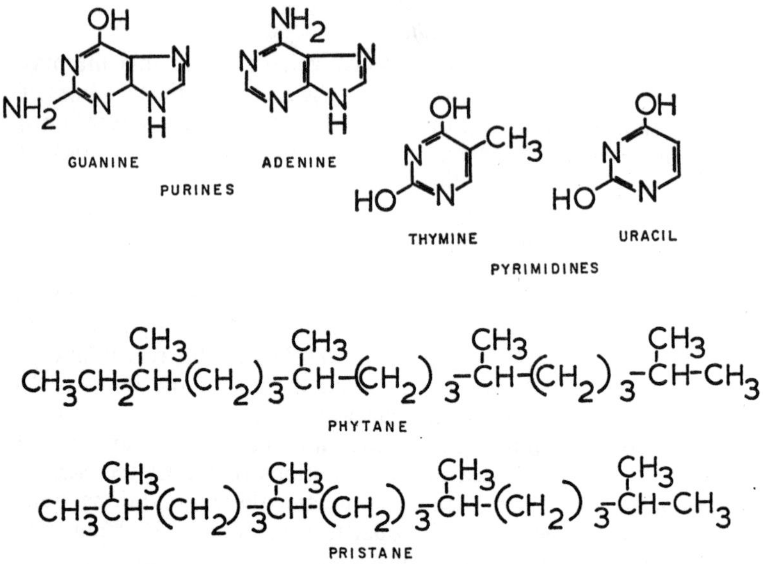

Fig. 45 The purine and pyrimidine building blocks of chlorophyll and two of its breakdown products, phytane and pristane, found in carbonaceous meteorites.

It is with these carbonaceous meteorites that we will now be concerned, since it is in these that we find some evidence for the existence of extraterrestrial life. A chemical study of these bodies has revealed the presence of several organic compounds which are the same as those found in early pre-Cambrian rocks on Earth – and, curiously, are also present in almost the same proportions. The most important compounds that have been discovered are purines and pyrimidines, which are the building blocks of the nucleic acids that are essential for life, and the hydrocarbons phytane and pristane, which are breakdown products of chlorophyll (Fig. 45).

We cannot, of course, be absolutely certain that these compounds were present in the meteorites before they entered the atmosphere. The possibility that they are merely contaminants introduced after they reached the Earth's surface is one which it is difficult to discount. Some ten years ago, this problem was reinvestigated by Duchesne and his colleagues at the University of Liège, using the modern technique of electron spin resonance. This method enables the presence of organic free radicals (molecules which have lost an electron, thereby acquiring a positive electrical charge) to be determined within meteorites. Positive results have been obtained in several cases but what is more important is that these free radicals are apparently distributed throughout the bulk of the meteorites, thereby proving that they are not due to contamination after these bodies have reached the surface of the Earth. Furthermore, it has been possible to show that the signals which have been detected are not due to organic matter which has been charred by exposure to high temperatures in a vacuum. From these studies, it now seems fairly well established that these molecules were present as such within the meteorites while they were out in interplanetary space.

Origin of Carbonaceous Meteorites

At first sight it may appear superfluous to consider whether the evidence just given is sufficient to indicate that life exists, or did exist, beyond the Earth, for the mere fact that we know these bodies enter our atmosphere from interplanetary space should be proof of this. The position, however, is not quite as straightforward as this. Although the meteors, the small particles of matter which burn out long before they reach the Earth's surface, are generally associated with comets, the meteorites are thought to be fragments of a parent planet, the remnants of which are the asteroids, small bodies which for the most part lie between the orbits of Mars and Jupiter. Whether this unknown planet exploded, or whether it simply did not form initially out of the cloud of dust and gas from which the solar system was born, does not concern us here. What is important is that, if they did originate in this way, it is clear that they are of extraterrestrial origin. Urey, however, has recently proposed a different theory, namely, that the meteorites originate on the lunar surface, being ejected into space by bombardment with comets. Now it is quite possible that at the time the Earth and Moon were formed they were much closer together than they are now; this being so, the composition of the outer portion of the Moon will be very similar to that of the Earth at that period. Indeed, we might almost say that the organic material we now find inside meteorites originally came from the Earth and is now being returned to us by this process of cometary bombardment many thousands of millions of year later!

This theory has not, however, been confirmed by analyses of the lunar surface made by the recent Surveyor landings, which have, instead, supported earlier work on the reflective properties of the surface of the Moon as a whole, indicating that the lunar crust is composed of basaltic material, quite different from that of the carbonaceous meteorites. A similar conclusion has been reached by laboratory examination of the rock samples brought back by the Apollo 11 mission.

At the present time most of the available evidence points to the fact that organic compounds which are normally associated with the presence of life are widely scattered throughout the solar system and that life can evolve, and in all probability has evolved, to a certain degree elsewhere than on Earth, the kind of life depending on the conditions prevailing upon different planetary surfaces.

Life beyond the Solar System

We must now extend the foregoing arguments and examine on a far wider scale the question whether there may be other planetary systems revolving around other suns than ours. The whole question of extra-solar planets arose quite naturally once it was established that the stars are merely suns, a large number of them very similar in all respects to our own.

At this point we must recognize that even in the most favourable cases of giant planets, much larger even than Jupiter, accompanying stars that are quite close to us, it is utterly impossible for any terrestrial astronomer to see or photograph them, irrespective of the size of the telescope he is using. Even in the largest instruments the stars are merely points of light in spite of their great size, and a body of planetary dimensions is not only much smaller but also will be very much less bright, since it will shine only by reflected light. From measurements made of the planets in our own system we know that they are far from being efficient reflectors of sunlight.

How then is it possible with our present instrumentation to show that such planets exist? To answer this question let us first take a look at the Earth–Moon system. Although we normally regard the Moon as revolving about the Earth, this is an over-simplification; in reality the two bodies are revolving about their common centre of gravity and the Earth traces out a very small orbit about this centre at the same time as it moves around the Sun in its elliptical orbit.

Starting from this principle we can see that, if a star has a planet (or planets) revolving about it, we might expect them to reveal their presence, provided that they are sufficiently massive, by their effect upon the motion of the star which, like the Earth, would appear to oscillate about a mean position as it moves against the background stars. From the angular displacement and its irregularities (Fig. 46) we would then be able to determine the mass of such a perturbing body and its motion. In addition, provided we know the distance of such a system, we could also derive its dimensions.

As one may imagine, these angular displacements will be extremely

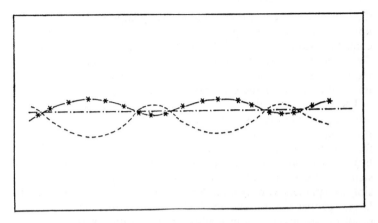

Fig. 46 The sinuous proper motion of a star which indicates the presence of an invisible companion that may be a planet.

small and difficult to measure. Nonetheless, a number of such measurements were made as long ago as the middle of the last century in the course of double-star studies. The first to be carried out resulted in the discovery of the companion of Sirius. We know that all of the stars are moving in random directions relative to the Sun, with a wide range of velocities. Visually, these motions are observed as the proper motion upon the more distant background stars, this being the component of a star's motion across our line of sight, motions towards or away from us being detectable only spectroscopically.

In the case of a single star that is, relatively speaking, close to us, the proper motion will be a straight line since there will be no other force acting upon it. In the case of double or multiple systems, the perturbing effect of the companions (which may be totally invisible) will come into play and although the centre of gravity of such systems will follow a straight-line course, the star itself will be found to move in a sinuous path.

In 1844, Bessel investigated the irregularities in the proper motion of Sirius and suggested that they were due to the presence of an invisible companion. Seven years later, Peters calculated the orbit of this body, which was eventually discovered by Alvan Clark in 1862 in almost exactly the predicted position. A second prediction by Bessel was also verified in 1869 when Schaeberle found the companion of Procyon.

Both of these small companion stars are white dwarfs of approximately the same size as the Earth but with masses similar to that of

255

the Sun. As a result of their comparatively large masses, their effect upon the motion of the primaries is relatively noticeable. When we come to consider a body of planetary mass, however, the problem of detecting and measuring such minute displacements from rectilinear motion is far greater. For this reason, it was long thought to be impossible to prove the existence of such planets.

The past two or three decades, however, have seen tremendous advances in photographic techniques, with a resulting increase in precision and several of the nearby stars are being closely examined for evidence of planetary systems. Van de Kamp has recently confirmed that Barnard's star, a 9·5 magnitude red dwarf in the constellation Ophiuchus at a distance of 5·9 light-years, has one, possibly two planets with sizes similar to that of Jupiter. This star has the largest proper motion known (10·3″ per year) and consequently the small deviations are more readily measured than with other stars.

Two other stars which may have planetary companions are 70 Ophiuchi and 61 Cygni. Both are visual binaries with periods of revolution of 88 and 720 years respectively. 70 Ophiuchi lies at a distance of 16·9 light years and microscopic examinations of photographs taken since 1910 enabled Strand and van de Kamp to calculate the probable mass of the invisible third body in this system. Unfortunately there is, as yet, no way of telling which of the two stars is accompanied by this small body. If it is revolving around the brighter of the two, its mass is about twelve times that of Jupiter; but if it accompanies the fainter star, its mass is even smaller, being not more than nine times that of Jupiter.

A similar conclusion has been reached for 61 Cygni. This visual pair lies only 11·2 light years away and a value of sixteen times the mass of Jupiter has been derived for this possible planetary companion. More work is, however, necessary before these cases are as fully authenticated as that for Barnard's star. There may also, of course, be other smaller and less massive bodies within these systems but at present it is not possible to detect them.

Other bodies of planetary mass have been suspected in the systems of ε Eridani, τ Ceti, and Ross 614, for example, and this leads us to a very interesting conclusion. All of these stars are very close to us and the mere fact that so many stars within a very small volume of space have been suspected of possessing planetary systems would indicate that the total number of such systems within the Galaxy as a whole must be very large. A conservative estimate has put the number at something like 100 million.

Thus the recent discoveries of these possible planets around other suns has completely reversed the view at the beginning of this century of the way in which planetary systems are formed. The old idea of a near collision of the Sun and another star had already fallen into disfavour before Strand and van de Kamp published their conclusions on 70 Ophiuchi and 61 Cygni. Later theories all suggest that planetary systems are formed far more frequently than was originally thought, being born out of the gas and dust clouds from which the central suns themselves evolved or from the debris of a companion sun which underwent a supernova explosion. We have already seen how infra-red studies of the T Tauri variables also strongly suggests that planets are being formed along with these protostars.

Life on other Planets

Having established that in our Galaxy alone many millions of stars are likely to possess families of planets, we must now consider the question of the suitability of a planet for the origin and sustenance of life. From what we know of the way in which life began and evolved on Earth, it would appear that the temperature of a life-bearing planet must lie within the range from 0 to 100°C, that is, the range within which water exists as a liquid. Outside of this range, it seems unlikely that life could begin. At lower temperatures any chemical reactions which might take place in ice would be far too slow for complex molecules to be built up, while at higher temperatures the chemical bonds between the carbon atoms would be completely disrupted. Such a planet must also possess a suitable atmosphere, which places a lower limit upon its mass; if it is too small the gravitational pull will be insufficient to hold any atmosphere and we would find, as in the case of the Moon and Mercury, that any gases which it may originally have had would soon dissipate into space.

Not only must the characteristics of the planet itself be like the Earth's, but the sun must also be of about the same size and temperature as our own. Very small, intrinsically faint stars will generate only enough heat for a planet revolving in an extremely restricted zone; suns that are too large, such as the red giants, or too hot, for example those of spectral classes earlier than F, would emit far too much of their radiation in the ultra-violet region of the spectrum for life to develop.

In spite of these somewhat stringent conditions, which we must impose if we take our own system as being typical, it is possible to calculate that there must be many millions of Earth-type planets within the Galaxy. On this basis, the chance that Man is alone in the universe is extremely small indeed.

Interstellar Communication

Once we accept the fact that the Galaxy may contain millions of Earth-type planets upon which life can exist, we must necessarily ask ourselves whether it is possible now, or at some time in the future, to communicate with intelligent beings other than ourselves. Even twenty or thirty years ago such an idea would have belonged to the realm of science fiction. Today, however, many astronomers are inclined to take it seriously. Already, we have taken our first step into space, though admittedly a very small step indeed. The technology for landing men upon Mars and Venus already exists and it seems highly probable that within the next two decades manned landings will have been made upon these planets.

The question of interstellar travel, on the other hand, is one that must clearly be deferred for a century or so. Here the distances concerned are so vast that hundreds or thousands of years would be required for journeys to even the nearest stars. The theory of relativity would appear to set an upper limit to the velocity with which a material body may travel, namely that of light, and even should this velocity – or something approaching it – ever become feasible, it would still take almost nine years to make the round trip to the nearest star.

Leaving aside the possibility that other races may have developed methods of interstellar travel based upon principles as yet unknown to us, is there any other means by which we can communicate with life on extra-solar planets? Fortunately, there is. Since the construction of large radio telescopes, designed for receiving radio waves from space, the idea of transmitting similar waves and beaming them towards some of the nearer stars that appear likely to possess planetary systems has been transformed into fact. We are however restricted in the kind of 'message' that may be transmitted. The one universal language which should, theoretically, be understood by alien races is that of pure mathematics. Accordingly, radio signals have been sent out consisting of simple dots arranged, like Morse code, into mathematical sequences. So far, of course, no reply has been received but when we take into account the distances involved, although such signals travel at the velocity of light, we must not be too discouraged.

Again, we must keep in mind the fact that the chances of any intelligent race having a technology similar to our own existing at the present time within a reasonable distance of the Sun is fairly remote. The period during which we ourselves have possessed the technology for transmitting and receiving radio signals is incredibly short

258

compared with the existence of life upon this planet, and even if we were to assume that within a radius of, say fifty light years of the Sun, there are a hundred other planetary systems in which life has appeared in some form or other, the chance that any of these are at this time in a similar state of scientific development is very small indeed. We may liken it to the chance that on a hundred columns each as high as Cleopatra's needle two segments only a hundredth of an inch thick are at exactly the same height. One might argue, of course, that once a technology capable of transmitting information over interstellar distances has been achieved, this ability will remain for centuries, possibly millennia. This, however, overlooks the fact that as scientific knowledge advances other means of communication will be developed, methods that will depend upon principles not yet discovered by us and therefore beyond our present capabilities to utilize. A negative result to our efforts must not, therefore, be construed as positive evidence against the presence of intelligent life elsewhere within the Galaxy.

15. Big Bang or Steady State?

So far in this book we have been concerned mainly with the discrete components that go to make up the universe as a whole, only briefly mentioning the large-scale aspects of this totality of all things. The problems with which we shall now be associated are inevitably quite different in many respects from those previously considered and it is for this reason that the overall study of the universe – its dimensions, age, evolution and particularly its origin and future characteristics – has been given a special name to distinguish it from the other, somewhat narrower, concepts of astronomy. We call this particular branch of the science cosmology. Until recently, though some of the best scientific minds of the past two centuries have contributed to its better understanding, cosmology was more of a speculative than an exact science, for astronomers were working with a paucity of data. All of this is now changing rapidly. Like all other branches of science, cosmology is based upon observational or experimental data and since, as we shall see, it bears a somewhat closer relationship to physics than to astronomy, it must also obey the known physical laws. Any speculations we may make concerning the universe as a whole, therefore, should be verifiable in the physics laboratory.

Now this statement at once raises a very interesting point. Some theorists have argued that here we are dealing with space, time and matter on such a vast scale (one theory even suggests that the universe is infinite on the spatial scale and eternal in duration), and that the laws of physics as we know them, although applicable to relatively small parts of the universe, do not apply to it as a whole.

This is a view that we must consider with some caution. The evidence we have at the moment would certainly tend to imply that most of the physical laws may be applied to the universe as we know it now.

Physical Constants

Considering the fact that the universe is on a vastly larger scale than our own Galaxy, or even the local cluster of galaxies, how can we justify the statement just made? We know, of course, that the stars we are able to see in other galaxies are in no way different from those in our own, that they are born, evolve and die in just the same way as the Sun was born, is evolving and will eventually die. Gravitation, too, plays the same role within the furthermost galaxies as it does here on Earth. Nevertheless, the distance and time scales that concern us when discussing the evolution of the universe are so large that we cannot dismiss out of hand the possibility that certain of these physical constants may change, either with time or position.

Let us start with an examination of some of these constants. Over a period of only a century, the rotation of the Galaxy, which we have already discussed, enables astronomers and physicists to sample quite a large portion of the local region of the universe, amounting to a volume of about 1 cubic parsec. If, therefore, any physical constant should change either with time or with position, detection of such a change, even though it will be extremely minute, should be within the scope of accurate measurements.

At the present time, such tests have been carried out to determine any such change in several physical constants with respect to others and all have proved completely negative. For example, Reines has shown that protons are very stable particles, having a lifetime of more than 10^{26} years, which is far longer than the estimated age of the universe on the basis of the Big-Bang theory. The alternative steady-state theory postulates that the universe is of infinite duration and we shall discuss this separately later.

Other constants which have been investigated include the electrical charge on the electron, which Hughes has shown to remain constant for velocities up to a tenth of that of light, and the neutral nature of ordinary atoms. In a neutral atom we find that the number of electrons equals the number of protons in the nucleus and again Hughes has proved that if there exists any difference between the charge on the electrons and that present in the nucleus it must be less than one part in 10^{21}.

One very important universal constant is that of gravitation. The crucial question of whether the gravitational constant changes with time is one that has not yet been fully answered. Dicke and his colleagues at Princeton University have carried out several experiments to show that over a period of a year any variation in this constant when measured against the electromagnetic interaction is certainly less than one part in 10^9. In the near future it seems possible that further refinement of these experiments will settle this particular question once and for all.

It is appropriate here, too, to consider the constancy of the velocity of light in a vacuum, since this is one of the basic principles of Einstein's general theory of relativity, which has a great bearing on the various theories of the origin and subsequent evolution of the universe. We know that the velocity of light varies in different media, the early measurements being made in air. More recently, very accurate measurements have been made *in vacuo* and certain investigators have reported small changes in this fundamental velocity. When examined objectively, however, none of these have been really substantiated.

Now where does all of this lead us? Clearly it shows that, at least as far as our own cosmic neighbourhood is concerned, the laws of physics apply throughout the entire region and there is no necessity for us to invent any new ones to explain the observations that have so far been made. We must now extend these arguments to encompass the universe as a whole and see whether they still hold. Since Hubble established beyond doubt that the galaxies are huge external star systems similar to our own and situated at enormous distances from us, our conception of the universe has been radically altered and extended from the fairly small, localized region accepted by most of the nineteenth-century astronomers to one of unprecedented distances, measured in hundreds of millions of parsecs. Out to the very limit of the largest optical and radio telescopes, the galaxies stretch away into space until at the present time more than a hundred million of them are known. As far as we are able to tell there is little, if any, thinning out of their numbers with distance and the same laws apply to them as to our own system.

The Expanding Universe

One of the fundamental problems which dominated cosmology for more than a century was that first put forward by Olbers as long ago as 1826. In essence, the question posed by Olbers was simple: Why is the night sky dark? At first sight this might appear to be quite trivial and scarcely worthy of serious consideration, but if we look deeper into the matter we find that, far from being so, it has a very great significance. As we know, the universe is populated with galaxies, more strictly with clusters of galaxies, as far as our instruments can probe out into space. Now if we assume that the universe is composed of an infinite number of concentric spheres such that the difference in the radius between one and the next is constant, then the volume contained within each of these shells will be proportional to the square of the radius. If we further assume that these clusters of galaxies are scattered uniformly throughout the universe (and at present all of the available evidence points to this being the case), the total number of stars within each shell must also be proportional to the square of the radius. The intensity of the light received from these stars is, however, proportional to the inverse square of the radius and accordingly the total radiation received at the centre from each shell will be the same no matter how far out we go, since one factor exactly nullifies the other. Consequently, for an infinite universe, the light at the centre will be so great that the entire sky will be as bright as the sun!

Since this is evidently not the case, it would appear that some important factor has been overlooked in our argument. The question is: What can it be? The first suggestion was that our Galaxy was the sum total of everything in the universe – simply a large, but finite, agglomeration of stars completely isolated in space with nothing but utter emptiness surrounding it, although it was agreed that the emptiness might stretch to an infinite distance in all directions.

With the recognition of the true nature of the galaxies as individual stellar systems, many of them being comparable in size and stellar content to our own and at incredibly remote distances, this view was no longer tenable. A second means of escaping from this dilemma was provided with the idea that, although the universe may be unlimited in space, there might still be a time limit imposed upon it. If, for example, the universe did not come into being until some 9,000 million years ago, then it would be impossible for light from any stars more than 9,000 million light-years away to have reached us. In other words, the furthest distance we could see, irrespective of the size of the instruments used, would be equal, in light-years, to the age in years of the oldest stars in the universe.

Now there is certainly some observational evidence in support of this idea. We have already seen that the age of the oldest stars in the Galaxy is about 7×10^9 years and, from what we know of the stellar content of other nearby galaxies, the same sort of figure applies there. Recently, however, Sandage has determined the age of certain star clusters and for one in particular, NGC 188, has calculated an age of between 13 and 17×10^9 years. This discrepancy would appear to be due more to errors in interpretation than observation.

Our next step must be to consider an extremely important discovery made some forty years ago by Slipher, Hubble and Humason.

The Red-Shift

As we have seen, the majority of stars have what is known as an absorption spectrum, one in which the bright continuum is crossed by many dark lines. There are, of course, some very hot and bright stars which lie high on the main sequence and give an emission spectrum consisting of a few bright lines on a dark background, but these are in the minority. The dark absorption lines just referred to are produced by the radiation from the hot interior of the star, the photosphere, passing through the cooler gases of the stellar atmosphere which absorbs those wavelengths corresponding to the particular chemical elements present in the atmosphere. When we come to compare the positions of these absorption lines with those of the same elements in the laboratory, we find that they are shifted slightly, towards either the red or the violet end of the spectrum. This shifting of the spectral lines is known as the Doppler effect. If the star is moving towards us, the lines are shifted towards the violet, while if the movement is away from us, there is a corresponding shift towards the red, the amount by which the lines are displaced being directly proportional to the velocity of approach or recession. We therefore have here a powerful means of measuring the speed with which a star is either approaching or receding from us.

Since the majority of stars within the galaxies produce absorption lines, the overall effect is an absorption spectrum. Indeed, this was one of the first clues to indicate that the galaxies are stellar systems and not large luminous gas clouds within our own Galaxy. The important discovery made by Slipher, Hubble and Humason was that, if this method is applied to the galaxies rather than to individual stars within our own system, all of those galaxies lying beyond the Local Group are found to be receding from us. There is not a single exception to this rule. Not only this, but the more distant a galaxy is, the greater is its red-shift and consequently the more rapid is its velocity of recession. In other words, the universe is expanding. The fact that, as far as we are able to probe into space using the largest telescopes, the relation between velocity of recession and distance is linear is well brought out in Fig. 47. This linear rate of expansion occurs in all directions, but we must not assume from this that our small Local Group of galaxies is situated at the centre of the universe. We hold no such privileged position. An observer in one of the more distant galaxies would find exactly the same relationship.

This expansion of the universe at once raises some very interesting points. Does the straight-line relationship depicted in Fig. 47 continue indefinitely or does the slope of the line alter owing to either a deceleration or an acceleration at great distances beyond the reach of present spectroscopic techniques? If the uniform rate of increase continues, will we eventually reach a point – at about 2,000 million parsecs – where the velocity of recession is equal to that of light itself? If the universe is, as we have already stated, the sum total of all things, *what is it expanding into?*

264

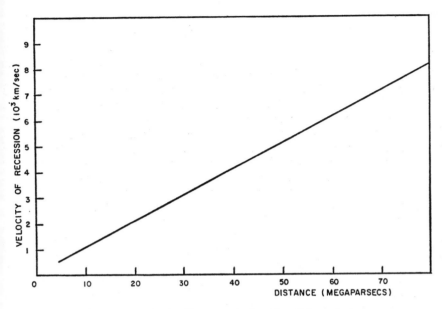

Fig. 47 The velocity-distance relationship of the galaxies.

With the aid of the 200-inch reflector at Mount Palomar, it has been possible to measure the velocity of recession of a cluster of faint nebulae more than 400 million parsecs away, these galaxies moving away from us at almost 40,000 miles every second (see Plate XVIII). Many more galaxies even more distant are known (shown in Plate XVII) but unfortunately these are so faint that it is difficult to obtain satisfactory spectra from which to measure their red-shift. So we can only say that out to this distance the expansion of the universe is essentially linear. How far we are justified in assuming that this relationship holds at greater distances is still a matter of conjecture. Certainly many astronomers are of the opinion that this linearity of expansion is a fundamental feature of the universe, one which was impressed upon it at the time of its creation. If this is so, then the answer to the second question posed above must be in the affirmative: there is an optical boundary to the universe, that at which the galaxies are receding with the velocity of light. Beyond this horizon, although there may well be other galaxies, it will be quite impossible for us ever to observe them since their light will never be able to reach us!

As far as the third question is concerned, this is a point that has

265

been generally ignored by astronomers, mainly because it takes us deep into the realm of pure speculation; any conclusions that may be reached depend entirely upon the researcher's own personal ideas and are quite incapable of subjection to any laboratory tests or observation. Nevertheless, it is a pertinent question and deserving of some kind of consideration. Clearly if we look upon the universe as the whole of a four-dimensional space-time continuum, as envisaged by Einstein and others, then whatever it is expanding into cannot be space-time as we know it. Perhaps, if any answer is ever to be obtained, it will come from pure mathematics rather than from astronomy or physics. Mathematicians are used to dealing with abstract dimensions higher than the fourth. The layman can fairly readily visualize the three dimensions of space, length, breadth and height, each being at right angles to the other two. When we come to introduce time as the fourth dimension, the mathematician may stipulate that we must regard this as also being at right angles to each of the three spatial dimensions, a concept that the layman will find difficult, if not impossible, to visualize. A simpler approach to the problem, perhaps, is to imagine any object, for example a cube, whose three spatial dimensions can be easily delineated. Now nothing material can possess zero duration, for this would mean that the object has no real existence at all. Therefore, in order to describe it completely, we must define its duration as well as its dimensions in space. The duration may be only of the order of a few millionths of a second, as in the case of certain unstable particles, or it may be as long as several thousand million years, as we find for the stars or the universe itself. Suppose now we were to add a fifth dimension. Certainly we cannot visualize the result of this, for the human brain is strictly limited in what it can conceive, but the result may possibly be the non-space into which the universe is expanding. This is really as far as we can go on this subject at the present time, for such arguments inevitably carry us no further than the stage of speculative suggestion.

The Big-Bang Theory of the Origin of the Universe

Once this linear expansion of the universe was recognized, it became apparent that if we were to move back in time we would find the fleeing galaxies much closer together than they are now. Indeed, if the expansion has always been linear it should be possible to calculate when all of the matter within the universe was conglomerated into an extremely small radius, this being theoretically the origin of the universe. Naturally this 'super-atom', as it was termed by Lemaître, one of the first exponents of this idea, would possess

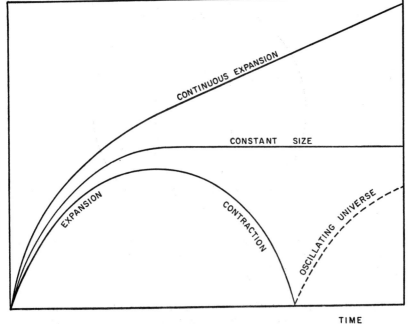

Fig. 48 Diagram illustrating the different types of universe depending upon the magnitude of the deceleration parameter.

an extremely high temperature and density. It would also be highly unstable and would immediately begin to expand. This theory of the origin of the universe is based mainly upon Einstein's general theory of relativity which, like Newton's mechanics, assigns a very particular property to gravity, namely, that the trajectories of all bodies in a gravitational field which have the same initial velocity and position are identical.

Einstein's theory is, of course, extremely complicated, containing several complex equations and conditions, and we might expect it to provide us with a large number of alternative theories of the origin and subsequent evolution of the universe. Curiously, this turns out not to be the case. If we assume that matter is uniformly distributed throughout space, then the general theory of relativity provides only one type of cosmology, that which was initially examined by Friedmann in 1922.

The special features of the Einstein–Friedmann theory are that

267

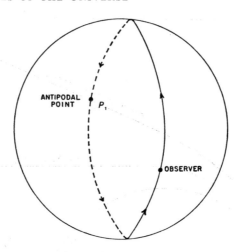

Fig. 49 The closed-universe model. Theoretically light from an object can travel all the way around the universe along a great circle. In practice, objects at P_1 are so distant that their light can never reach us.

the universe is in constant motion and that it began initially with zero radius (Fig. 48). It must be made clear at the outset that the idea of a universe with zero radius is one which many astronomers are loath to accept, especially as the density would then be infinitely high. Mathematically, this situation is known as 'singularity', simply meaning that, although we can apply the well-known laws of physics up to this point, we can go no further. At the very instant that the universe began expanding, the physical laws came into operation. Before that, they did not exist. We shall have more to say on this point later when we come to discuss the steady-state theory which does not require the existence of such a singularity.

Now although this theory implies that the universe will expand, it does not clearly differentiate between two possibilities. One is that there will, at some time, come a deceleration, a slowing down of the rate of expansion. The universe will reach a certain critical radius and then, since we know that a static universe, whether it be of a very small size or extremely large is unstable, it will contract back to zero size. Such a model is known as a closed universe (Fig. 49). Alternatively, the expansion could go on indefinitely, with the galaxies getting further and further apart and the mean density falling steadily. In this case we have an open universe (Fig. 50).

268

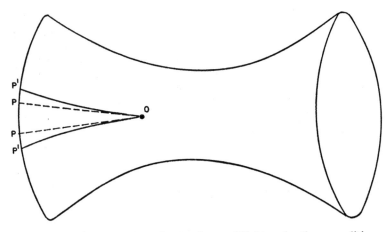

Fig. 50 Model of an open universe. A ray of light under these conditions will continue to travel outward and can never return to its starting point. Light rays will follow the curved paths OP^1 and not the straight paths OP.

Whether the universe is open or closed depends upon two factors: the amount of matter it contains and its energy content. From the data which have been accumulated, it would appear that if the density is greater than 3×10^{-29} grammes per cubic centimetre the universe will be closed, and if less than this it will be open. At first sight it would therefore seem to be a relatively simple matter to decide between these two models. Unfortunately, this is far from being the case. The average density of matter within the galaxies has been estimated as 7×10^{-31} grammes per cubic centimetre, about one-fiftieth of that necessary for the closed model of the universe to be correct. Does this necessarily mean that we are forced to abandon the closed model? The answer is no, for the simple reason that the matter we find in the galaxies (stars, gas and dust) is only a part of the total within the universe. There exist other forms of energy and matter which are extremely difficult both to detect and to measure accurately. Until we are able to do this we cannot be certain which of the two models is correct.

Ghost Particles

One major source of universal energy which has recently been discovered is that associated with the neutrinos. We have already made the acquaintance of these particles when we discussed the evolution of the supernovae. Since they are so important in the present context it is now necessary to consider them in some detail.

We have already seen that ordinary hydrogen, the simplest of all the chemical elements and also the most abundant, consists of a central proton and an electron revolving in an orbit around it, the opposing electrical charges on these two particles exactly neutralizing each other. Suppose now we examine the element carbon. Here we find six electrons in the cloud around the nucleus and consequently there must be six protons in the nucleus. Since the electrons have a negligible mass (it would take 1,847 electrons to equal the mass of a single proton), we would expect the mass of the carbon atom to be six times that of hydrogen. In actual fact, the value is slightly greater than twelve. The excess mass comes from six other particles inside the nucleus known as neutrons, particles which have a mass almost identical with that of a proton but no electrical charge. Other carbon atoms are known which contain seven and eight neutrons in the nucleus and which have different atomic weights, being known as isotopes. We cannot, however, continue adding neutrons to the nucleus without limit. If we were to have an atom of carbon with an atomic weight of fifteen (six protons and nine neutrons) we would find that it is highly unstable – the disparity between the numbers of protons and of neutrons is now too great – and one of the neutrons within the nucleus will turn into a proton, an electron being ejected in the process. Since there are now seven protons in the atom, it is no longer an atom of carbon but one of nitrogen. We already know of this type of atomic transmutation, for we have encountered it in the familiar carbon-nitrogen cycle which is predominant in stars with masses somewhat higher than that of the sun; it is the well-known β-decay process.

Now one peculiarity about the β-decay process is that the electrons which are ejected from the nucleus do not all have the same velocity, even though the states of the nucleus at the beginning and end of the process are the same for all carbon atoms which undergo this decay transformation. We know that the velocity of the electron depends upon its energy; the more energetic it is, the higher will be its velocity, and vice versa. As it turns out, there is a second particle emitted with the electron, this being the neutrino. The total amount of energy released during the β-decay of the nucleus remains constant, but the proportions of it associated with the electron and the neutrino can vary within quite wide limits.

The neutrino possesses neither mass nor charge and is consequently extremely difficult to detect and identify. However, it may soon become possible to determine the intensity of neutrino showers reaching the Earth from the Sun and from interstellar space. Because of the two properties mentioned above, the neutrino is the most penetrating of all the known particles, passing through hundreds of feet of rock without any hindrance. Accordingly, since almost every other kind of particle will be absorbed completely by such a depth of earth and rock, experiments are being carried out in deep mines, although even here the number of neutrinos that can be detected is extremely small.

It has been discovered that the neutrino takes part in one nuclear reaction with an isotope of chlorine producing an atom of argon, an inert gas which can be readily identified and accurately measured. The method used is to fill a large tank with an organic chlorine-containing compound such as carbon tetrachloride or tetrachloro-ethane (ordinary cleaning fluid), the tank being situated at the bottom of a deep mine. The incoming neutrino will then pass through the liquid, reacting as it does so, with the formation of argon which is then drawn off and measured, giving an indication of the number of neutrinos that have passed through the tank in a given time.

As far as cosmology is concerned, the neutrinos are produced inside the stars and, since they take no further part in the usual nuclear reactions, they pass out into space taking a certain amount of energy with them. In most stars the principal mode of energy loss is by radiation but, as we have seen, there are some, particularly those stars which eventually become supernovae, which lose so much energy by the escape of neutrinos from their interiors that they collapse utterly. The number of neutrinos within the universe as a whole is therefore far from negligible and they must be taken seriously into account in any attempt to derive a model for the universe.

Mass and Energy Content of the Universe

At this point it is instructive to examine the total energy and mass within the two different models of the universe according to the Einstein–Friedmann cosmology. In the closed model, the total energy and mass is finite. The universe begins with zero radius and immediately starts to expand and will continue to do so until a critical point is reached. Just what is this critical point and how is it determined? The explosive creation of the universe imparted the energy to it which caused it to expand, and for a very long period (several thousand million years) this is the dominant motion in the universe. Once stars and galaxies form, however, gravity becomes more prominent, and with time the expansion of the universe slows down as gravitational attraction between the galaxies becomes more important. A time will therefore be reached when gravitational attraction equals the explosive motion imparted to the universe at

271

its creation. When this occurs, expansion ceases and we have a static universe. This, however, is just as unstable as was the initial state and, as gravitational attraction is now the dominant force, the universe commences to contract, eventually returning to its original state with zero radius.

What happens next is something of a problem for cosmologists. Some believe that a further expansion will then occur and that the universe will continue expanding and contracting endlessly. On this basis it is impossible to tell whether we are now existing in the first expansion of the universe, the hundredth, or even the ten-thousand-millionth. Other astronomers are of the opinion that there will be one, and only one, expansion and contraction. If we turn to Einstein's theory of gravitation for help in this problem we find that, as far as it is able to tell us, the latter case is the more probable. On the other hand, we must admit that Einstein's equations are not valid for the initial conditions of the universe when it has zero radius. The whole question is still therefore very much an open one.

Before we discuss the open model of the universe, it will be helpful here to mention one particular constant in the Einstein–Friedmann equations known as the deceleration parameter, q_0. This is simply an expression of the rate at which the expansion of the universe is slowing down. The family of curves shown in Fig. 51 are those for different values of q_0 and, as we see there, if it is exactly 0·5 then the universe will expand to a certain size and will then continue for ever at this size. On the other hand, if the value of q_0 is greater than 0·5, the density of matter is sufficiently high (greater than 3×10^{-29} grammes per cubic centimetre) to eventually stop the expansion and allow contraction to begin. Suppose, however, that q_0 is less than 0·5. In this case the universe will continue to expand indefinitely and we have an open model. This simply means that as time goes on the density of matter within the universe grows steadily less and, even after all of the stars in all of the galaxies have burnt themselves out, expansion still continues, although the universe of matter is, to all intents and purposes, completely dead. Even though all of the stars have now expended their energy, however, there must still be energy present within this dead universe to enable it to go on expanding.

The Red-Shift–Apparent-Magnitude Relation

Earlier in the chapter we considered the linear nature of the red-shift–distance relation which led us to the conclusion that the universe is expanding. We must now return to this in order to see whether it will help us to distinguish between the various models of the universe that have just been described. Then we saw that, as far

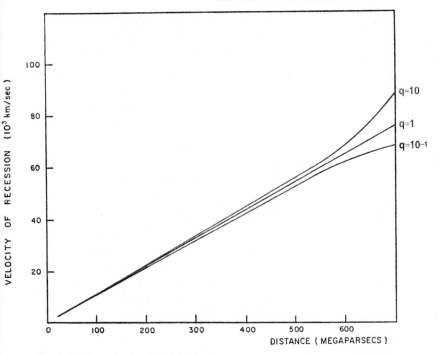

Fig. 51 Deviation from the linear velocity-distance relationship with different values for the deceleration parameter q0. At present it is difficult to observe objects sufficiently distant to determine whether there is any real deviation.

as measurements have been made, the red-shift is directly proportional to the distance of the galaxies, the latter being determined by their apparent magnitude. We also stated that it was generally believed that this linear relation held out to even greater distances than can be measured at present.

With the introduction of the deceleration parameter, we must now modify this statement a little. Fig. 51 illustrates how this apparently linear relation will be slightly changed, depending upon the value we give to q_0.

There is quite an appreciable scatter among the points derived from the astronomical data we have at the moment and, until techniques are developed that will enable us to extend these observations further, it will not be possible to decide which cosmological model best fits the observed facts.

The difficulty lies not in the determination of the apparent

S   273 magnitudes (from which we derive the distances), for with the 200-inch reflector it is possible to record galaxies as faint as magnitude 23 by means of long-exposure photography (see Plate XVII). Rather it is the fact that, although there is sufficient light from these faint objects to affect the photographic plate, there is simply not enough good spectra to be obtained from which the red-shift may be measured. At present it is almost impossible to obtain such spectra for objects much fainter than magnitude 17.

The Evolution of the Einstein-Friedmann Universe

As we have seen, the Einstein–Friedmann universe begins with a singularity which is highly unstable so that expansion begins. Within 0·01 seconds after creation the density has dropped slightly to about that prevailing within the nucleus of an atom, while the temperature is around 200,000,000,000°K. At this stage, any particles that are produced, including protons, are extremely unstable and react with one another to produce a host of peculiar particles, very few of which are still in existence. Most of these are what are known as anti-particles, which means that they are identical to those particles with which we are now familiar except that their electrical charges are reversed. An anti-proton, for example, is simply an ordinary proton in every way except that it has a negative charge instead of a positive one.

As the expansion continues, the temperature of both matter and radiation decreases quite rapidly. One second after creation, the temperature is 15,000,000,000°K and all of the anti-particles have been annihilated with the exception of the anti-electron (the positron); we are left with electrons, positrons, neutrons and neutrinos. It is at this stage that a very significant reaction occurs, the result of which is now, after all these millions of years, providing astronomers with some confirmatory evidence of the Big-Bang formation of the universe. Electrons and their anti-counterparts, the positrons, react together to form high-energy radiation at an incredibly high temperature. We shall discuss the fate of this primordial radiation a little later.

Over the next twenty minutes or so after creation, with the temperature falling to around 400,000,000°K, a gradual build-up of elements occurs, the result at the end of this period being a conversion of about a quarter of the hydrogen into helium. At this temperature, however, most of the atoms present will be in the ionized state, that is, with some of their outer electrons missing. Because of this, they will be positively charged since the number of negative charges (electrons) is now less than the number of positive charges (protons)

274

in the nucleus. Not until the temperature of the universe has fallen to a few thousand degrees Kelvin will these ionized atoms capture enough of the wandering electrons to render them electrically neutral. All of this time, the energy of radiation within the universe will greatly exceed that of matter. Gradually, however, with increasing expansion and the accompanying decrease in temperature, the intensity of radiation energy falls below that of matter and at around 300°K (normal room temperature), the globular clusters and then the galaxies begin to form, followed by individual stars. From this point on, we have a matter universe as opposed to the initial one containing primarily radiation. At the present time, the average temperature of the universe has fallen as low as 3°K, slightly above absolute zero.

The Steady-State Universe

As we saw earlier, not all astronomers subscribe to the Einstein–Friedmann theory of the evolution of the universe. We must now consider the reasons for this and an alternative theory, which has a more recent origin than that just described. One of the major difficulties associated with the Big-Bang theory is that the origin of the universe cannot be described by the laws of physics. We have already seen that Einstein's general theory of relativity, upon which it is based, is a classical theory and cannot be used to describe conditions at the moment of creation. It is probably true to say that any theory which contains a singularity such as this is viewed by scientists with suspicion. At the very moment of creation, all of the matter and energy within the universe was packed tightly into what is virtually a point source, where both temperature and density are infinitely high. No one doubts that such a condition is highly unstable and that under such circumstances expansion must occur. The difficulty as far as some astronomers are concerned is that several salient features of the universe appear to have been imparted to it at the moment of creation. We cannot question the way in which these featured were produced and, if we were to ask why the universe came into being as it did and what came before this singularity, no answer could be given. All that can be said is that, if certain conditions were impressed upon the universe at the moment of its creation (conditions that appear to some to be quite arbitrary), then the universe would evolve as it did. In other words, the Big-Bang origin is not subject to either physical or mathematical analysis.

It was for this reason that Hoyle, Bondi and Gold put forward the idea of a steady-state universe, one which avoids the necessity of postulating a singular origin by substituting a universe in which

275

matter is being created all the time. Now how does this overcome the difficulty of having a superdense state to begin with? Simply by arguing that all of the matter which is now present in the universe did not exist 9,000 million years ago! In the same way, 9,000 million years from now, very little of the present matter in the universe will be observable owing to dispersion. If, instead of creation's being a singular event, we postulate that it is continuous, then the universe always has been in existence and will continue to be so. In this way the singularity is completely removed.

We can also examine the physical properties that such a universe will have. First of all, it will expand in the same way as the Einstein–Friedmann universe but not for the same reason. Here it is the constant formation of matter within the universe which forces it to expand, and it is this rate of creation of matter that in turn controls the rate of expansion. As time goes on and the galaxies move further apart, new galaxies are formed so that, in contrast to the Einstein–Friedmann universe, the density of matter remains constant. The crucial point is, of course, the mechanism by which matter and energy is created. Certainly it is well known that mass can be transformed into energy (this is exactly what happens in the atomic bomb or a nuclear reactor). Conversely, whenever a photon, a unit of light energy, is transmuted into an electron and a positron, energy is converted into mass. We find a similar situation in the β-decay process where an electron is created when a neutron changes into a proton. However, these are all very small-scale creations of mass. Although this may well be a means by which matter may be created on a scale large enough to comply with the requirements of the steady-state theory, we do not yet fully understand the physical laws behind this creation of mass and energy. This does not mean that they do not exist. only that so far they have not been worked out satisfactorily.

Hoyle has suggested that there is a creation field acting within the universe in a similar way to the gravitational field that acts between the stars or the galaxies. This creation field, however, exists over even greater distances than that of gravity and since it has such a universal influence it explains why matter is so evenly distributed throughout the universe, because its influence is determined by the furthermost corners of all existence.

We now have one very sharp distinction between these two opposing theories. The Einstein–Friedmann idea of the universe states that all of the galaxies were formed about 6,000 million years ago and they are now all of about the same age. On the basis of the steady-state theory the galaxies are of all ages from the youngest to the oldest and theoretically we ought to be able to utilize this distinction to differentiate between them.

How can we do this? By making use of the finite velocity of light. Whenever we observe the stars or the galaxies we are seeing them, not as they are now, but as they were at some time in the past depending upon their distance. The light we see tonight from α Centauri left that star just over four years ago. If it were suddenly to turn nova at this moment we would not know of the event for another four years or so. Similarly, if the whole of the Andromeda nebula were to vanish it would be over 2 million years before we knew of it. In other words, we are seeing the Andromeda nebula as it was over 2 million years ago. When we extend our observations to more distant galaxies, we are actually looking even further back into time.

Now how does this help us to distinguish between the Big-Bang and the steady-state theories? If the Big-Bang idea is correct, then the further we probe into space, the younger will the galaxies at these distances appear to be, since we are seeing them as they were nearer the epoch of their creation. On the other hand, if the steady-state theory is correct we shall find the same mixture of young and old galaxies at very remote distances as we do within our local region of the universe.

This problem has been extensively investigated in recent years and, although it may eventually prove possible to obtain an unambiguous answer, the results so far are still far from concordant. The chief difficulty is that we are dealing with objects so remote that they lie at the very limit of our present instruments. Several of the large radio telescopes have been used in an attempt to resolve the problem since they are capable of probing much farther into space than optical instruments. The balance of evidence obtained from very distant radio sources appears to lie with the Einstein–Friedmann theory, although a great deal of work remains to be done concerning the actual nature of many of these sources.

The Primordial Radiation

One prediction of the Einstein–Friedmann model of the universe was, as we saw earlier, that the annihilation of electrons and positrons shortly after the creation led to the production of a high-temperature radiation which, over the millennia, has now cooled to about 3°K. The prediction of the presence of this radiation (although at a somewhat higher temperature) was made in 1946 by Gamow and his colleagues. Working from similar assumptions and using more recent data, Dicke predicted the value of 3°K. In 1961, Penzias and Wilson, working at the Bell Telephone Laboratories, measured the atmospheric microwave noise at a wavelength of 7 centimetres. This background noise diminishes slowly as one moves an antenna from the horizontal direction toward the zenith, but from their results they discovered a small, residual radiation which, at this particular wavelength, corresponded to a black-body radiation at 3°K, exactly that predicted by Dicke. In addition, the characteristics of this radiation are those we would expect from the primordial radiation on the basis of the Einstein–Friedmann universe. The most recent rocket observations, at a wavelength of about 1 millimetre, however, give the disturbing result that the intensity of the background radiation may be higher than expected for a temperature of 3°K.

16. Relativity

In the last chapter we discussed the various models of the universe and saw how most of them are based, in one way or another, upon Einstein's restricted and general theories of relativity. To complete our discussion of the universe, therefore, we must finally consider in some detail the two theories of relativity and particularly how they have been applied to cosmology.

Astronomy, and indeed almost all branches of science, have progressed on two broad fronts; first there comes the discovery of new facts, and then the discovery of formulations or systems to account for the facts already known. As might be expected, the most outstanding landmarks in science have proved to be of the second kind. Undoubtedly this has arisen from the human desire to do more than merely accumulate and catalogue facts: to search for theories to explain such facts and, where the theory is in conflict with observation, to modify the theory.

As far as astronomy is concerned, this process of selection and modification has gone on for more than two thousand years and relativity is only one of the latest theories which has been used to explain the observed behaviour of certain aspects of cosmology in particular. The early notion that the planets revolved about the Earth in perfect circles, together with the Sun and fixed stars, led to the idea that the retrograde motions occasionally observed in the movements of the planets were due to their moving in secondary circles about the primary ones. This mechanism of cycles and epicycles held the field until in the sixteenth century Tycho Brahe proved that Mars deviated from its calculated position by as much as eight minutes of arc. During the following century, the whole edifice of astronomy that had been built up at the time of the early Greeks was demolished. The Earth was removed from its privileged position as the centre of the universe. Kepler substituted elliptical orbits for the planets in place of circles and the underlying force of gravity determined the planetary motions with an unprecedented accuracy.

The concept of gravity as put forward by Newton, however, decreed that, so long as the perturbing influence of the Sun and the other planets were taken into account, the ellipses along which each planet travels should repeat themselves exactly. Again, the motion of

279

one of the planets – Mercury – was found to be in disagreement with the Newtonian idea of gravity when Leverrier discovered that its orbit is rotating in its own plane at a rate of forty-three seconds of arc every century. At the time it seemed inconceivable that the Newtonian concept was wrong and various investigations were made to discover some unknown mass that could be producing this discrepancy. Among the hypothetical masses that were investigated were an unknown planet revolving within the orbit of Mercury, certain unsuspected density differences within the Sun itself, and a band of diffuse matter lying in the equatorial plane of the Sun. In each case, however, the mass required to produce the observed effect upon Mercury was so great that not only would it have been readily discovered but it would also have produced a similar, although probably smaller, effect upon the other planets. The true explanation of this anomalous rotation of Mercury's orbit did not come until Einstein showed in 1915 that the Newtonian idea of gravity had to be slightly modified to comply with the more recent ideas of space and time.

Towards the end of the nineteenth century, it was becoming obvious to physicists that the classical theory put forward by Newton was at variance with many of the observations that were being made, and that a fundamental rethinking of the older concepts of time and space was necessary. Newton had stated as one of his laws of motion that it is impossible to determine any point of absolute rest from a study of the motions of the Sun, the stars or any other body in our region of the universe. Since the publication of his main work, the *Principia Mathematica*, the idea that there might be an all-pervading medium, known as the ether, existing throughout the whole of the universe had grown steadily. It was further believed that this medium was necessary for the propagation of electromagnetic phenomena, that light, for example, from the most distant stars travelled through the ether in the form of waves. Furthermore, the ether had to be at absolute rest since, it it were distorted in any way by the presence of massive bodies lying within it, this would not agree with the astronomical observations made on the aberration of light.

If such an ether existed, then physicists and astronomers were presented with a method for measuring the absolute velocity of the Earth. Putting it another way, all one had to do was measure the velocity of the ether moving past the Earth.

The Michelson-Morley Experiment

In 1887, the American physicist Michelson made the first attempt to measure this velocity using the known velocity of light. Now

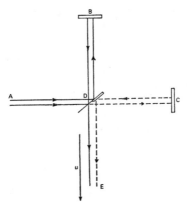

Fig. 52 The Michelson–Morley experiment which proves the non-existence of an all-pervading ether throughout the universe.

although it is not possible to make a direct measurement of the velocity of light along a single straight course, it can be done by reflecting a beam of light back along the same path by means of a mirror, and Michelson's experiment was based upon this simple property. If the Earth is travelling through the ether with a velocity u, then the velocity of a beam of light through the ether relative to a terrestrial observer should be $c-u$ when sent in the direction of the Earth's motion and $c+u$ when sent in the opposite direction, c being the velocity of light in the ether.

Fig. 52 shows in diagrammatic form the apparatus used by Michelson to measure the difference in the velocity of light beams sent in two directions at right angles. The beam of light from A was split into two beams at D, these being at right angles to each other. The arms BD and CD were made as nearly equal in length as possible and the whole apparatus was mounted so that it could be rotated about the point D. If we assume that the Earth is moving in the direction DE, the times taken for the two beams to travel along their respective arms and back to D will be different. There will, quite naturally, be some inherent difference due to inequalities in the lengths of the arms but if the whole apparatus is rotated slowly about D these differences will vary owing to the motion of the Earth. In spite of the accuracy of his apparatus, however, Michelson found it impossible to detect any difference at all. Light appeared to possess exactly the same velocity no matter in which direction it was transmitted. Working in conjunction with Morley, Michelson refined the equipment until finally it should have been possible to measure any variation due to the motion of the Earth through the ether of as little as 0·5 kilometres per second. To have accepted this result as it stands would have been unthinkable since it would imply that the Earth is at absolute rest, taking us back to the geocentric idea of the universe.

In 1893 and 1895, FitzGerald and Lorentz independently suggested that one way out of this dilemma was to suppose that every moving body shortens in the direction of its motion, this Lorentz – Fitz-Gerald contraction, as it is called, being given by the equation

$$L = L_0 \sqrt{1 - \frac{u^2}{c^2}}$$

where L_0 is the length of the body when at rest, L its length when it is in motion, u its velocity and c the velocity of light. This contraction is exactly that required to nullify the Michelson–Morley experiment and it has led to some very astonishing but perfectly logical conclusions. Firstly, no material body can travel at or in excess of the velocity of light; otherwise its length in the direction of motion becomes either zero or imaginary. Secondly, the velocity of light is the same for every observer irrespective of his own velocity.

This is in reality the principle of relativity as put forward in 1904 by Poincaré. More broadly, he stated that all of the phenomena of nature will be the same to any two observers who are travelling with any uniform velocity relative to each other. The following year, Einstein brought out his restricted theory of relativity, which amplified this hypothesis and also tested it against the established physical laws. In every case relativity has been found to be in accordance with known physical laws and, in those few cases where it has appeared to be at variance with any law, further experiments have vindicated relativity and shown the laws themselves to have been erroneous.

The Constancy of the Velocity of Light

The idea that the velocity of light *in vacuo* is the same for every observer regardless of his own velocity has had an extremely important bearing upon our ideas of space and time. Since this has also had a profound influence upon cosmology, we must examine these changes from the older Newtonian idea in some detail.

Let us suppose that we have two observers each equipped with a clock which they wish to synchronize with each other and that they are also capable of sending and receiving a radio signal. If observer A

282

sends out a signal when his own clock registers exactly midnight, observer B must make an allowance for the time taken for the signal to reach him, the time of receipt of the signal at B being given by

$$T = T_0 + \frac{x}{c}, \qquad (2)$$

where T_0 is the time when A sent the signal, x is the distance between the observers and c is, as usual, the velocity of light. This is the method that astronomers normally use to synchronize their clocks. However, if the Earth is moving in the direction of B with a velocity u, then the velocity of the signal will not be c but $c+u$ and the two clocks will not be synchronized, but will differ by the interval:

$$x \left(\frac{1}{c} - \frac{1}{c+u} \right) \approx \frac{ux}{c^2} \qquad (3)$$

The value on the right-hand side of expression (3) is only approximate but it is sufficiently accurate for our present purposes and it clearly shows that it is impossible, according to the principle of relatively, to synchronize the two clocks, since we can never find the value of u.

Let us take another example which will further illustrate the impossibility of two observers' being able to compare the simultaneity of an event. Suppose that two observers A and B are moving past each other at a high velocity and that at the instant they pass a light flare goes off between them. As we have seen, the velocity of light is not affected by either the speed of the observer or that of the source. Consequently, A and B will each be able to show that the light is travelling outward in all directions with the same velocity, and by means of a suitable arrangement of mirrors each can further show that the light of the flare is in the form of a sphere with himself at the centre. Observer A will use his coordinates x, y and z along his three spatial axes to define the points of the sphere and a time coordinate t for the time that has elapsed since the flare went off. Similarly B will use x', y', z' and t' as his co-ordinates. If each observer then measures the position of the wavefront at any given time, he will do so by means of the well-known equations:

$$x^2 + y^2 + z^2 - c^2 t^2 = 0 \qquad (4)$$
$$x'^2 + y'^2 + z'^2 - c^2 t'^2 = 0 \qquad (5)$$

Prior to 1905, a mathematician would have connected the two series of co-ordinates by the following simple relation

$$x' = x - ut, \, y = y, \, z' = z, \text{ and } t' = t \qquad (6)$$

Even a cursory glance will show that equation (4) cannot be transformed into (5) by means of these relations and it was left to Einstein to show that this can only be done if A's co-ordinates are connected to those of B by the relations:

$$x' = \frac{(x-ut)}{\sqrt{1-u^2/c^2}}$$
$$y' = y \tag{7}$$
$$z' = z$$
$$t' = \frac{(t-ux/c^2)}{\sqrt{1-u^2/c^2}}$$

An inspection of the relation given in (6) and (7) will reveal that two basic alternations have been made: the FitzGerald contraction factor has been introduced into the equation of the x-axis to allow for the change in length due to the motion through the ether with a velocity u, and the time axis has been modified by the same factor so that the velocity of propagation of light along the x-axis may still have the same velocity c. The time equation has also been modified by the inclusion of the synchronization factor (3).

These transformations had, indeed, been used by Lorentz in 1895 to show that all electromagnetic phenomena are the same for an observer at rest in the ether as for one moving through it with a velocity u. Consequently, they are generally known as the Lorentz transformations. Relativity is conceptually different for Einstein showed that they hold for any two observers whose relative motion to each other is u and as a result there is no need for postulating the presence of the ether. In general terms, if two observers are moving towards or away from each other at velocities that are an appreciable fraction of that of light, they will make different measurements of distance, time and mass, and all of these measurements are equally true.

The mass of a rapidly-moving object will increase with its velocity, whereas its time scale will run slower. Both of these results have important and far-reaching consequences in physics and astronomy. Several methods are available to prove both of these concepts quite rigorously. The masses of elementary particles have been shown to increase as their velocities approach that of light; atomic accelerators are used to boost the velocities of these particles. In addition, certain unstable, short-lived species have been found to remain stable longer if they are moving rapidly than if they are at rest. Such particles also slow down the emission of their radiation, with a corresponding shift to the red of their spectral lines. This particular

Doppler shift has to be taken into account when we measure the distances of galaxies that are receding from us – owing to the expansion of the universe – with high velocities. This is clearly demonstrated by the radio galaxy 3C 295 which, according to the shift in its spectral lines, is apparently receding from us at about half the speed of light. The restricted theory of relativity, however, taking into account the amount by which the red-shift has been enchanced by its recession, yields a real velocity of recession of only about a third of the speed of light.

The Four-Dimensional Continuum

If we wish to study the motion of any body in space, we have to measure four quantities, namely, the three spatial co-ordinates and a time co-ordinate. Unfortunately, it is not possible to construct a four-dimensional graph but we may visualize this four-dimensional continuum as being something real, the history of the moving body being represented as a continuous line, termed the world-line by Minkowski. On the basis of relativity, every observer will define the same world-line for any moving body within the same four-dimensional continuum, only their choice of axes being different owing to their various relative velocities. For example, a terrestrial observer may calculate that the supernova of 1054 actually exploded 5,000 years before the birth of Christ, whereas an observer in another part of the Galaxy may reach a somewhat different conclusion regarding the occurrence of the two events. Both estimates will be equally true, although not absolutely so. Thus the concepts of space and time are merely subjective ones and only the four-dimensional continuum has an independent existence. This is basically the condition envisaged by Minkowski.

Relativity and Gravitation

The restricted theory of relativity considered the measurements that two observers make whenever they are travelling through the space-time continuum in straight lines and with constant velocities. Although this satisfactorily explained the propagation of light and other electromagnetic phenomena, it did not give any really satisfactory explanation of gravity which, on the old Newtonian hypothesis, is inconsistent with relativity.

Einstein therefore extended this earlier theory into the general theory of relativity, which relates the measurements made by observers who are moving in curved paths and are also accelerating relative to each other. One of the conclusions reached was that a gravitational field is quite indistinguishable from an inertial field due to acceleration or a change in direction. Now the one thing that will create an intertial field is matter itself. The Sun produces such a force upon the Earth and all of the planets, just as the Earth does upon the Moon, and vice versa. Accordingly, Eistein was forced to the conclusion that the presence of matter in the universe always produces a curved field which not only forces such bodies to form into spheres but also makes satellites of such bodies move in either circles or ellipses. The only geometrical properties that space posses- ses are those which are impressed upon it by the presence of matter. Not only this, but the degree by which space is curved is directly related to the mass of the bodies concerned.

All of this led him to the realization that space does not obey the simple straight-line theorems of Euclidean geometry. The shortest distance between any two points in space is not a straight line, but a curve, just as it is on the surface of the Earth. Fortunately, the mathematics of multi-dimensional curved geometry had already been worked out some seventy years earlier by Riemann, who had extended the mathematics of curves into any number of dimensions.

As far as gravitation is concerned, there are three main predictions made by the general theory that admit of experimental verification. The first is one that we have already mentioned, namely, the strange discrepancy in the perihelion motion of Mercury which observa- tionally amount to 43 seconds of arc every year. From the known mass of the Sun, Einstein was able to calculate the degree of curvature of the surrounding space this mass would produce and its effect upon Mercury. This is $42 \cdot 9$ seconds of arc per year, in excellent agreement with the observed value. There is, of course, a similar effect upon all of the other planets but here the test is by no means as stringent as in the case of Mercury, since the predicted motions are extremely small and precise measurement is extremely difficult.

The Bending of Light by Gravity

For centuries it has been assumed that light travels in a straight line between the source and the observer, such a line being known as a geodesic. In general relativity, a geodesic is curved owing, as we shall see later, to the overall curvature of the universe. Theoretically, too, it should be curved whenever it passes close to any massive body which, as we have seen, distorts the continuum in its neighbourhood. As a result, a ray of light will travel in what is, to all intents and pur- pose, a straight line only when it is in interstellar or intergalactic space, away from any disturbing body. Any curvature impressed upon it by the curvature of space itself will be extremely small. When it passes close to a body such as the Sun, however, the degree

286

of bending should be within the capabilities of measurement. From the general theory, Einstein predicted that a ray of light just grazing the surface of the sun should be bent by 1·745 seconds of arc, the amount of bending becoming gradually less in proportion to the inverse distance from the centre of the Sun.

In 1919, astronomers from Greenwich and Cambridge observed the total eclipse of the Sun in order to test this prediction, photographing the stars close to the Sun during totality and comparing their positions with those they have when the Sun is not present at that spot. This method has been used on subsequent occasions by various teams of astronomers. The results obtained are mostly very close to the predicted amount, although certain discrepancies have been recorded. The difficulty appears to lie in the fact that those stars for which direct measurements have been obtained lie several solar diameters from the disc of the Sun, the figure for the bending of light at the Sun's surface being obtained by extrapolation and this, of necessity, introduces some uncertainty.

The Gravitational Reddening of Light

The general theory of relativity also predicts that the spectral lines of any atom should be displaced towards the red (this should not be confused with the well-known Doppler shift, which we have already discussed in detail elsewhere). As far as the Sun is concerned, this displacement is quite small, amounting to about 0·008 Å unit at a wavelength of 3883 which is the wavelength due to the cyanogen molecule at which most observations have been made. Unfortunately this small displacement is masked by other effects and early measurements were very discordant although later ones, particularly those of Evershed, are in good agreement.

Since the magnitude of this displacement increases with the mass of the emitting body, the best chance of observing it accurately will be in the case of extremely massive stars. Certain difficulties arise, however, if we attempt to use supergiant stars since the large radii of these stars mean that the magnitude of the shift is only about that found in the case of the Sun. The white dwarfs, however, are quite a different proposition. Here, a mass similar to that of the Sun is compressed into a volume not much larger than that of the Earth and measurements of the gravitational red-shift of the companion of Sirius – a typical white-dwarf star – by Adams using the 100-inch reflector yielded a result of 0·32 Å, compared with the theoretical value of 0·30 Å. There is, therefore, no longer any doubt of the validity of this spectral shift.

Relativity and Cosmology

Until the restricted theory of relativity was put forward in 1905, the two concepts of space and time were extremely simple ones, namely, that the only property possessed by space was that of extension and by time that of duration. On this basis, every element of space has the same nature and it must necessarily be infinite in extent in all directions if it is to follow the rules of Euclidean geometry. Similarly, time must be eternal in duration. A little reflection will show why this must be so. If space were to stop at any particular point, then the last element of space must be different from all of the others, which contradicts the statement we have just made. In the same way, if time were to end at any instant, this final instant of time would be different from all of those which had gone before.

This view was altered drastically by relativity. Space and time were replaced by a space-time continuum which, in the neighbourhood of matter, is curved to a degree depending upon the mass of the disturbing body, while out in the depths of the universe, away from all matter, it is not. Once it was accepted that all elements of space-time are not intrinsically the same, the infinity of space and the eternity of time were no longer necessary concepts for the description of the universe. A curved universe that is finite and yet unbounded may then be postulated according to the terms of general relativity, but whether it is positively or negatively curved is a problem that can be answered only by observation. The most recent investigations of the furthermost galaxies indicate that the universe is positively curved, but they are still too inexact to admit of a definite answer.

Mention has already been made of the different models put forward to describe the universe. Here we may consider them in more detail. In 1917, Einstein added a further constant to the equations describing a gravitational field – the cosmical constant. This factor represents a repulsion effect within the universe which exactly balances the mutual gravitational attraction of the stars and galaxies. The particular model of the universe envisaged by Einstein was one in which the presence of matter brought about a closed system, with the radius of curvature and the density of matter remaining constant and the cosmical constant counterbalancing the gravitational attraction. The Einstein model is essentially a static one, but in 1930 Eddington proved that such a universe is unstable and will expand. Eventually, a limit will be reached when all of the matter contained within such a universe is so widely separated that any effect which individual bodies have upon each other is vanishingly small. Such a model had already been described by de Sitter, this being one that does not vary with time (although test particles will drift apart due to expansion), and is virtually empty of matter. Other models have been described by Lemaître and Friedmann, both of these being non-static universes in that they do vary with time, and both intermediate between the Einstein and the de Sitter models. Indeed, it may be said that the Einstein universe progresses through the Lemaître and Friedmann types as it tends towards the de Sitter universe.

The Radius of the Universe

Evidently the universe is extremely large even if it is positively curved. Nevertheless, it is possible to determine its radius once we know the rate of expansion, by means of a formula derived by Lemaître on the basis of the fact that, if x is the separation of two galaxies at any given time t, then the value of $(1/x)$ (dx/dt) tends towards a limit as the expansion progresses, this limit being

$$\frac{c}{R\sqrt{3}} \tag{8}$$

where c is the velocity of light and R is the radius of curvature in the Einstein universe. Now the value (dx/dt) is simply the rate of recession due to the expansion of the universe, a factor that can be measured observationally, and combining this with the limit (8), we obtain a value of 10^{24} kilometres for the radius of the universe.

The Mass of the Universe

A second characteristic of the universe that may be calculated according to relativity is the total mass of matter in space. This is done by means of a formula derived from the mathematics of the Einstein universe, namely,

$$\gamma M = \frac{\pi c^2 R}{2} \tag{9}$$

where γ is the constant of gravitation as put forward by Newton and c is again the velocity of light. Substitution of the various constants into this equation yields a value of 10^{52} kilogrammes for the total mass of the universe. These figures have their full meaning only if applied to either the Big-Bang or the oscillating models of the evolution of the universe. On the basis of the steady-state theory they are valid only as characteristic of that part of the universe we can observe, for here the space-time continuum is regarded as having no beginning and no end as regards either extent or duration

289

and, since matter is considered to be continually created, the total mass of the universe is virtually infinite.

We have come a long way from our first exploration into the boundaries of the universe, into the remarkable achievements of the astronomers and astrophysicists, particularly those carried out over the past half century or so. From a study of the lunar surface at a mere quarter of a million miles from the Earth we have progressed outward to ever-increasing distances until finally we have reached the dim, nebulous no-man's land which lies on the very rim of that which we can observe. In the light of all the discoveries that have been made, it is perhaps instructive to recognize that once we pass beyond the bounds of our tiny solar system all of the information which has been obtained has been patiently extracted from the feeble radiation that reaches us from the distant stars and galaxies.

Inevitably, certain topics have received only a brief mention and there are others that it is impossible to enter into too deeply within the framework of this book. Nevertheless, it is hoped that the reader has been given an insight into the problems that still face the modern astronomers in his attempt to understand the working of the universe and an indication of the lines along which research into this science will be carried in the future. As was stated at the outset, astronomy is perhaps the most dynamic of the physical sciences, and the recent discovery of the quasars and pulsars in particular will inevitably lead us to a revision, possibly of a somewhat drastic nature, of many of the theories held at present.

Index

291